A Strategy for Using Multicriteria Analysis in Decision-Making

Nolberto Munier

A Strategy for Using Multicriteria Analysis in Decision-Making

A Guide for Simple and Complex Environmental Projects

Springer

Nolberto Munier
Carrer Poeta Josep Cervera i
Grifol 14 - 33
46013 Valencia
Spain
nolmunier@yahoo.com

ISBN 978-94-007-1511-0 e-ISBN 978-94-007-1512-7
DOI 10.1007/978-94-007-1512-7
Springer Dordrecht Heidelberg London New York

Library of Congress Control Number: 2011933229

Printed on acid-free paper

Springer is part of Springer Science+Business Media (www.springer.com)

Preface

First of all, it is necessary to point out that this is not a mathematical book, albeit obviously the subject of decision-making rests a good deal on this discipline, since it is almost impossible to take decisions without the help of some indicators, ratios, weights, procedures, algorithms, etc. which are in essence mathematical concepts. However, *bien entendu*, it is the Decision Maker (DM) who takes a decision, not a mathematical model, no matter how elaborate or sophisticated it could be. Once this is clarified, perhaps the reader asks what the purpose of this book is then. It aims at giving the DM the information he/she needs to collect the necessary data; to analyze that intelligence and facts, to process them, and to extract valid conclusions.

However, because usually the decision-making process is a complex task, with large amounts of information, it is extremely difficult or even impossible to take a rational decision, due to the number of intervening variables, their interrelationships, potential solutions that might exist, diverse objectives envisioned for a project, etc.; therefore some help is called for, and some strategy is required to organize, classify, and evaluate this information. The strategy includes methods to analyze the results and to get help from them. This book aims precisely at developing such strategy.

Help comes in the form of computerized models or methodologies which not only take the burden off normally tedious calculations (which also offer the opportunity for many mistakes), but, which is more important, are designed to obtain coherent responses based on a rational analysis, on the grounds of a series of principles and algorithms particular to each method. The DM may or may not accept the result yielded by such models, whatever they might be and whatsoever the reasons, but they offer an outcome which could be analyzed by the DM, adding or deleting concepts, alternatives, criteria, etc., and from this point of view this book tries to be a guide or road map.

This work reviews the main different approaches devised for helping the decision-making process, with rather complicated examples posed and solved for each one, and using, when available, dedicated software. However, within the strategy, this book emphasizes the use of a methodology called Linear Programming (LP), because there are compelling grounds for this choice, which are based on the optimality of results that it produces and for its mathematical soundness, as well as because it offers a less subjective approach to a complex problem.

Brief Description of Each Chapter

Chapter 1 is devoted to understanding the decision-making process and its complexity, defining the objectives of a project, and its constituent elements such as criteria, thresholds, and alternative scores on some predetermined scale. It also covers aspects such as the way to gather information, analyzing data and identifying impacts that any project generates. This chapter suggests ways to support the DM, not only in analysis of the information but also in selection of the methodology used. It helps the DM with information for his/her orientation about decision-making examples on actual cases and in different areas, giving the name of the project, its author/s and electronic address. Its purpose is to support the DM in the first stages of the process by providing an opportunity to contact another DM on similar projects. Why this? Because decision-making is not an exact science even when based on healthy mathematical principles; it intervenes in so many areas of human activity that it is often rewarding to have a look at what other people have done when faced with similar problems.

Chapter 2 refers to collection and processing data. Needless to say, this is a very important subject, because the final outcome of the decision-making process will depend on the quality of data and information imputed to a model. There also exists the matter of subjective information, which is always present, especially in aspects related to the effects of a given project as it impacts the population and the environment. For this reason, we emphasize the need to allow for as much people participation as possible because, more often than not, citizens provide important information that was not initially considered by the DM and his/her staff. Regarding this subject, the chapter reveals some procedures about different methodologies to evaluate impacts, including the SWOT analysis.

The chapter also analyzes different types of projects regarding their interrelationship – another important feature – as well as different criteria types – a fundamental and usually misunderstood concept – and threshold standards for different criteria types. Finally, it treats impact evaluation – another typically neglected or grey area – and offers a graph methodology for this task. As a bottom line, this chapter provides fundamental information to guarantee, as much as possible, a correct set-up of the problem and paving the way for modeling its solution.

Chapter 3 includes a review of the most usual methods for decision-making which are based on different postulations for selection, such as the utility theory (MAUT), the outranking procedure (ELECTRE and PROMETHEE), establishing preferences (AHP and ANP), and those grounded on the concept of analyzing distances to the best solution, such as TOPSIS. In each case a brief and concise explanation of each model is given, and then a real-life illustrative example is proposed and solved 'by hand' to illustrate the different steps required by each methodology.

However, and since most actual cases are too complex to be solved this way – and also with little effectiveness due to calculation time and errors – they are worked out using dedicated computer packages. For this reason three real-life cases are put forward and solved by the respective software, that is, using *'Decision Lab'* for PROMETHEE, (Project: *Route selection for an oil pipeline*); employing *'Super Decisions'* for AHP, (Project: *Location analysis for a factory*); and utilizing *'Six Sigma'* for TOPSIS, (Project: *Selection of urban alternative roads*). Each result is analyzed for the special possibilities that each program offers for sensitivity analysis, and comments are made on each model and their pros and cons.

Ultimately, this chapter's purpose is to make the DM and his/her staff conscious of the characteristics of diverse models or of computer programs available for decision-making. Since any given project is unlike any others due to its own unique features, and also because each DM thinks differently, this chapter tries to give enough information for the DM to decide which is the most appropriate method to use, according to his/her liking and the characteristics of the project. No model is considered to be better than another, therefore it cannot be said that a certain method will give more reliable information or results than another; in fact, usually the same problem yields different results according to the model used, which is explained by the different assumptions that distinguish one method from another.

In addition, not all projects are similar; some have very detailed and quantitative information with little subjectivity, such as in construction and manufacturing projects; others have a lot of subjectivity because of their nature, e.g. social projects and those affecting people's lives, such as a highway construction in an urban area. Finally – and this is the most usual – all projects have a blend of quantitative and qualitative information, a mixture of objective technical data as well as subjective appreciations such as land use and environmental issues; some have few uncertainties while others are the opposite, such as projects that rely on prices, sales, demand, etc. Of course, the DM's preference for one model or another will be at least partly based on his/her own experience.

Chapter 4 is devoted to Linear Programming (LP) as the technique was originally designed. It is a more complicated procedure than those analyzed in Chap. 3, also the oldest, and perhaps a little more difficult to understand because it is based in pure mathematics, specifically in matrix analysis. However, the DM and analysts do not need to go into the complexities of its theory to understand and apply the model, since it can be perfectly grasped with an example for an elementary case, which is done in this chapter, using an illustrative case of selecting between two sources of renewable energy, and solving it graphically.

The model works as do the others by simulating a real situation using a set of linear inequations and then fixing a linear objective. In our example all of these values can be represented in a plane (because there are only two alternatives) – that is in graphic form – and the solution immediately found. It has the advantage that the reader 'can see' the logic of the procedure and can easily understand the result. Once the solution is found, the analysis goes on to investigate how possible variations of some parameters can affect the result, also in graphic form, which is very important for sensitivity analysis, that is to answer the *'What if....?'* question. Naturally this is only an example whose main usefulness is that it 'can be seen'.

With real-life problems the model works exactly the same way, except that it cannot be visualized because, as each alternative is a dimension, in projects with more than three alternatives, options, or programs, it is impossible for us to imagine them, let alone to see the problem graphically. However this is not an inconvenience for linear programming which works in mathematical spaces – not in physical ones; because of that, it can admit hundreds and even thousands of alternatives and criteria, in any order or magnitude, as long as they are linear, which is something that no other model can do. And the most interesting point is that – provided a solution exists – the model will always find the best solution or optimum, another feat that no other model can match. An additional advantage is that there are many computer programs to solve this problem; one of them comes as the Excel® add-in used in this book.

At this point the reader will probably be wondering why such a remarkable tool is not used more often and in lieu of other techniques. The answer is simple; LP works in the way described only when there is one objective function, which is not usually the situation in most real-life cases, where there is not only one objective but several, some of them mutually contradictory. However, LP can be used in a manner to avoid that drawback, as described in Chap. 6.

Chapter 5 is devoted to analyzing in detail the kind of information that LP can provide. To do that, another actual example is proposed, this time a little more complex than the graphic example used in Chap. 4. For that purpose, a project with five objectives and three options or alternatives is set up, aimed at selecting alternatives for electricity generation in a region. The project is described, solved using LP, and its conclusions thoroughly discussed. Direct and indirect impacts are considered as well as thresholds for environmental contamination, electrical energy output, number of jobs created, etc.

The main instrument here is a pay-off matrix which allows us to determine the influence of running *ceteris paribus*[1] each objective at a time, following a thorough analysis of each scenario with a lot of information for the DM. This intelligence illustrates the importance of each objective and it can be so significant as to produce a change in the original ideas, plans, or thinking, as exemplified in this case. What is important in this analysis is the fact that there is room for potential confirmation or negation of original assumptions made by the DM and staff, and consequently the DM can not only understand the benefits or lack of them but – very important – *is able to justify now or in the future, why a certain decision regarding selection of a project has been taken.*

The chapter introduces the *'shadow prices'* concept, a unique feature of LP, and the base for sensitivity analysis, which is, by the way, mandatory in a well-constructed decision-making process. The chapter finishes with a simulation by posing some

[1]Ceteris paribus: Latin expression which means is *'Others things being equal'*. It is a concept widely used in economics when one wishes to determine the effect caused by a variable when all the other variables of the problem are held constant. That is, one value is chosen to be variable and all other values are held constant.

typical and relevant questions that the DM might ask, and for which the model has to have answers.

Chapter 6, recognizing that the principles of LP have been outlined in Chap. 4 and the analysis of results in Chap. 5, now deals with the SIMUS method, which reiteratively applies LP to solve a mono-objective problem using in turn different objectives. The result is expressed in a matrix whose elements or scores are optimal, and that allows one to reach a solution to the problem, although without guaranteeing optimality. To explain the method, another actual case is used, in this case '*An airport expansion plans*' with three alternatives and 20 criteria. It also presents a SIMUS variant using the same set of optimal scores matrix, but under another concept. There is complete agreement with the result reached by SIMUS compared with another methodology (REGIME), which was the way the original problem was solved. However, this could be just a coincidence since both models differ in their assumptions, and their correspondence is by no means certain. However, the fact that two completely different methods reach the same result appears to indicate that their result is the best.

This agreement with the results from other models, has been also verified by this author when comparing results from SIMUS with results on 45 actual projects, which are detailed in the Chap. 9, Table 9.9, where in a high proportion of cases there is complete agreement not only in the first selection, but in the still more difficult job of ranking the projects and considering many diverse models.

Chapter 7 aims to give more information to the DM and his/her staff since it makes comparisons between different methods considering diverse points of view. It does not try to determine if one is better than another; its intention is to enlighten the DM about the technical characteristics of each model. For instance, both ELECTRE and PROMETHEE methods offer a family of models, from which the DM will be able to select the version which best matches his/her needs or expectations. An example of this can be seen when the DM can choose the way the information is presented using PROMETHEE I or PROMETHEE II as illustrated in the case when analyzing this method in Chap. 3. Another similar comparison is made, but considering now the working characteristics of each model regarding its particular difficulties, available software, mixing projects, sensitivity analysis, etc. Its purpose is to illustrate for the DM the methods he/she can be most comfortable with, or the method that better conforms to his/her requirements. To round out the example, the sequential different steps that must be followed for each methodology implementation is detailed; in this way it attempts to give information about the difficulties and the execution time needed.

Chapter 8, in its first part, is devoted to *complex projects* and what is meant by that designation. There is much to say about this subject considering, among others matters, complexity, size of the problem, limiting features, as well as integration of effects, impacts and externalities. The chapter considers the structure of a complex project to be a matrix structure, similar to that used in many consulting firms. This perceived similarity rests on the fact that the different criteria types – which are almost always present – are most generally common to all projects, in the same manner as services from diverse departments of a consulting firm are shared by all projects.

Similarly, as a consulting company can manage several different projects at the same time, in decision-making there are several projects or alternatives that must be considered simultaneously.

The chapter continues with the presentation of a comprehensive set of complex real-life projects and with their analysis and solution by LP, although of course, all are abbreviated and condensed versions of the real cases, not only due to confidentiality but also because in a book like this there is no room for their complete analysis. Seven cases are presented as follows:

First case: *River basin planning*
Analysis of a very complex project solved by the Massachusetts Institute of Technology (M.I.T), some decades ago, to determine the best uses for a scarce resource, using water from a river in Argentina. It is a classic in the river basin literature.

Second case: *Selecting environmental indicators*
This difficult problem consisted in selecting a manageable set of environmental indicators out of a large number of them. The chosen indicators had to comply with an internationally known framework, and also obey tough restrictions regarding compliance of demands from a Canadian governmental agency, which required a minimum number of indicators per criteria. In this case, analysis is also done when there is incomplete data – a common case – as well as finding the best solution, that is, the best mix of indicators that maximizes the information provided by data, through the use of the entropy concept.

Third case: *Housing development*
This case analyzes the situation of a Mexican entrepreneur wishing to develop a housing complex. The selection consists in determining the number of houses of different types to be built, which is restricted by land use, size of each plot, minimum and maximum floor space, availability of municipal services, etc. Sensitivity analysis is also performed assuming some changes in the City Hall provision for some services, as well as answering some entrepreneur's questions about criteria relative importance, or how he can reduce his costs, or how his profits will be influenced by changes in some parameters.

Fourth case: *Municipal projects*
The objective in this case is to select a group of projects in order to make the best possible use of potentially available funds to minimize costs, in a large Argentine city. It is a complex problem because there are temporal as well as physical restrictions between projects, that is, there are projects that cannot be started unless another project has been finished. There are besides financial restrictions from the point of view of synchronization of percentage of estimated work done in a certain period with funds available in the same period. The project assumes at the beginning that there will be enough funds to execute them all, however, unexpected circumstances reduce the amount of available funds; thus, the model has to determine which is the best blend of projects that can be executed with this reduced funding.

Fifth case: *Scheduling bridge repairs*
Another very complex scenario, that took place in a semi-rural area in Canada, with many road bridges. The necessity to have a plan for bridge repairs in different time periods along with bridge conditions, costs to repair or replacement, together with availability of funds from different sources, made it a difficult case to determine when to repair eight bridges. The problem consisted in determining a schedule for repairs, since each bridge could be repaired immediately, in the short term, on in the long term, which made 24 options. On top of that in the first case there were three options regarding urgency. This case shows different areas where sensitivity analysis was performed regarding changes in the tentative schedule, financial terms varying in amount and timing, percentage of completion, etc.

Sixth case: *Land use and rehabilitation of abandoned land*
This case – common around the world – pertains to an American city that suffered the consequences of declining rail passenger traffic due to air competition, and a change in the philosophy of sea cargo due to switching to container ships, since both produced respectively abandoned railway stations and prime vacant land in the city from abandoned wharves and depots. Apart from the natural complications of this whole project which considered seven schemes subject to 15 criteria, some of them of qualitative nature, and mixing maximization (for instance in transportation), with minimization actions (such as environmental impacts), there were more restrictions, since some of the projects could be associated with others, that is, congruency existed between them. Therefore, the problem consisted in selecting which of the schemes should be developed in order to maximize an envelope of benefits for the city.

Seventh case: *Contractors' selection for metallurgical development*
An American consultant was chosen by a South American government agency to design a metallurgical complex for copper concentration, high in the Andes Range. The consultant had also to pick the general contractor company to perform the job using local companies. The process started with a shortlist of five companies, and then calling for bids, albeit restricted to the shortlisted firms. Terms specified that not only independent companies could bid but also joint ventures formed with this purpose. The selection was based on 15 qualitative and quantitative criteria on diverse areas, and the construction companies were weighted according to referrals from industry; however, after the first result was known, the Project Manager demanded a sensitivity analysis to verify what would have been the selection if all companies were given the same weight, which for him was more realistic, since he was not very confident in the referrals. The result showed a clear winner in both approaches and this was the chosen firm.

All these cases are solved using LP and it can then be appreciated how the tool can manage different and difficult situations.

Appendix

The Appendix has been developed as a support for the techniques commented on in different chapters. Two examples are proposed:

The ABC bike manufacturing company
This fictitious example has been designed to demonstrate the use of the S.W.O.T. analysis, acronym of Strength, Weakness, Opportunities and Threats. It is a simple example with the intention to show how to proceed to evaluate impacts.

The second case pertains to risk analysis and applies to '*Selecting construction alternatives for a subway line to traverse a river*'. It is a common project; however its importance lies in the fact that it considers risks, and these might be important enough to alter any plan. The proposed example corresponds to a real case in Spain, albeit values and criteria have been changed, and serves to illustrate how to identify and evaluate risks.

The final part of the Chap. 9 shows Table 9.9 where 66 different and actual cases are detailed. Each project gives information about its title, area to which it belongs, author/s name/s, sources where the corresponding paper or article can be found, and their electronic addresses. This table contemplates the following methods: AHP and ANP, ELECTRE, PROMETHEE, COST/BENEFIT, LP AND SIMUS, MAUT, REGIME, SAMI and some others. Its purpose is to support the DM by giving her/him the opportunity to study different approaches to a problem that could be similar.

The table includes projects for the following 39 areas (number between brackets indicate projects I.Ds. in the table).

Advertisement (65)
Agriculture and cattle (25-39-41)
Air Force (45)
Airports (02-03-10)
Commerce (63)
Communications (64)
Decision-making theory (47)
Electrical generation, distribution and strategies (11-12-26-66)
Energy policies (33)
Environment (34-36-40-44-57-62)
Forestry (48)
Government (22)
Housing development (16-32-52)
Industrial location (09–50)
Industry (46)
Investment alternatives (15)
Land use (28)
Location analysis (21-37)
Medicine (51)
Merchandise distribution (23)
Mining (27)

(continued)

(continued)
People recruiting (14)
Pipeline construction (24)
Purchasing (08)
Recycling (19-20-35)
Regional infrastructure (61)
Renewable energy (49)
River basin (13)
River waterways (01)
Roads infrastructure (04-05-07)
Sewage (18)
Solid waste (06)
Urban infrastructure (43)
Urban investments (42)
Urban rehabilitation (30-38-53)
Urban strategy (54-55-56-58-59-60)
Urban transportation (31)
Water distribution (17)

Contents

Chapter 1
Strategic Plan for Decision Making: Its Constituent Elements

Abstract Since this work refers to strategy, and considering a loose definition of the term, its intention is to establish a plan to reach an objective, that is to make the best possible selection of projects, alternatives, and programs. In this chapter, this plan follows a logical sequence of steps, with comments just to give the reader a clue or indication of the importance of each one. The whole idea is that, going through the chapter, a decision maker team may be aware of what they need before doing anything. In other words it attempts to give a scheme of the decision making process This general outline will, on the other hand, allow all members of the team to learn what is the responsibility of each one.

Keywords Strategy • Objective • Decision making • Criteria • Thresholds • Data

1.1 The Decision Making Process: The Challenge

When there is a set of projects, comparables or not, subject to restrictions and limitations, and when it is necessary to perform a selection and ranking, i.e. when there is a complex choice, it is convenient to use strategic planning to solve this system, that is to apply a set of generally sequential procedures, usually with a lot of feed back to find the most convenient result. The fact that normally one doesn't know what will be the best result that can be achieved makes the problem really difficult. That is, say that the problem is to select a city between several others in order to start a hardware manufacturing and technological centre.

If several cities are considered as candidates with similar attributes regarding population size, average technical education, universities, Information Technology history, etc., and one city is chosen, it is almost impossible to predict how the project will perform in the selected city compared with what would be its performance in other sites. Certainly, a very well defined goal can be set up, but it belongs to the future and as it is impossible to work in successive approximations we have

N. Munier, *A Strategy for Using Multicriteria Analysis in Decision-Making: A Guide for Simple and Complex Environmental Projects*, DOI 10.1007/978-94-007-1512-7_1,

to trust in our best judgment. Possibly the answer is in the future once the plant is in operation, but unfortunately (or fortunately) it is not allowed for us mortals to know it.

This brief introduction serves to illustrate the nature of the problem this book addresses. Simply put, a person or a group of persons, based on the available information on quantitative data, assuming values on intangible and subjective issues, identifying and estimating potential risks, and confronted most of the time with uncertainties, must decide what the most convenient solution between several options is. To this it is necessary to add that, in a large project, there could be multiple effects not only in the environment but also in the economy and in the life of generations to come. What a responsibility and a challenge!

Fortunately, we now have tools that can help, only help, in this difficult endeavor, and this is precisely the aim of this book: To develop a methodology to allow the DM[1] to make a reasonable, educated and documented decision. It is like having our ignorance organized.

1.2 Understanding the Importance of the Decision Making Process

This book develops a strategy for the decision-making process with emphasis on complex projects and suggests a methodology that will deal with the most important component in the process, the DM. In so doing, this strategy aims at supplying accurate, reliable and thorough information for the team to take a sound, facts-based, unbiased and documented decision. There is also a special emphasis on the environmental aspect because, no matter the kind of project considered, there are always environmental impacts, most of the time negative, which need evaluation and minimization.

This is no doubt a formidable task, for, besides its complexity, it carries a lot of responsibility, the future life of many people who will be affected by the selected project, and the use (or misuse) of billons of euros. Many of these projects will have a long life, surviving the lives of the people who projected, financed and built them, and may influence in many different aspects the geography of the planet. They can be responsible for the construction of new cities, such as the city of Las Vegas, because of the Boulder Dam. They can drastically affect communications from the construction of highways, high-speed trains or tunnels such as the Channel Tunnel. They can have an unpredictable technological impact like the construction of the Super Collider,[2] and affecting the lives of people living nearby, etc. Naturally, most of the projects are not of this magnitude, but the principles to select them are identical for them all.

[1] From now on the expression *'decision maker'* (DM) will be used to identify either a person or a group of persons who are in a position to render an educated opinion or judgment.

[2] Refers to the Large Hadron Collider built by the European Organization for Nuclear Research (CERN) on the border between Switzerland and France for high-energy experiments.

The strategy starts with the process of determining the envisioned *objective* or objectives. It continues with the *relationships between projects* or alternatives,[3] the *impacts* that a project can produce (not only in the environment but also in society and the economy), the selection *criteria* and corresponding *thresholds*, and goes on with the *analysis* of some techniques currently used to help the decision maker. Regarding these techniques it is common knowledge that *computer programs* exist, written with the purpose of helping the decision making process, albeit not for solving the problem, but *to shed light on the decision issue* where so many factors must be taken into account. This book deals with a methodology not very often used to work out problems with multiple objectives, called Linear Programming[4] (Dantzig 1948).

Why Linear Programming (LP)? Because this technique is the only one that guaranties optimal solutions; however, its drawback is that it works with just one objective; for that reason, in this strategy, a variation of LP called 'SIMUS' (acronym of **S**equential **I**nteractive **M**odel for **U**rban **S**ystems) is utilized.

SIMUS exploits the best qualities of LP, but at the same time, and recognizing that most projects aim at several different and sometimes contradicting objectives, develops a procedure to allow LP to also treat these multiobjective projects. This is also important because many times there is not too much interest in adopting the best mathematical solution that emerges from a model, but that one which best satisfies the objectives of the organization promoting a project. However, knowing which the best or more appropriate theoretical solution is may serve as a yardstick to appraise the one finally chosen.

One of the most relevant aspects of this strategy is that it deems the decision making process as a system and as such analyzes its different components, their forward and backward relationships as well as their organization, interrelations and feedback. The strategy encompasses the following areas:

1.3 Components of a Project

1.3.1 Objective

The objective of a project is to reach some pre-established goal/s; these goals can apply to many different things, as for instance the manufacturing of a product with costs as low as possible, or to determine the best location to install an industrial plant.

---

[3] From now on the expression 'projects', 'alternatives', 'programs', 'developments', 'plans' are used with the same meaning, that is undertakings of different kind, size and purposes, to be built, implemented, organized, for a purpose, objective, goal or target.

[4] Linear Programming: Very well-known mathematical procedure for solving linear problems subject to restrictions in a large variety of situations and also complex decision-making scenarios. Its most valuable and widely used application is due to the American mathematician George Dantzig, who in 1948 developed an efficient algorithm called '*Simplex*', used nowadays worldwide in hundreds of applications. This technique is considered by many as the most powerful mathematical tool invented in the twentieth century.

It also applies to ranking a set of projects according to their relative importance and aiming at the same goal, which can help decision-making. Or to minimize damages to the environment because of certain human actions, or maximize a certain government policy, or to determine the best route between two cities, which also applies to pipelines, transmission lines, etc. Consequently, it is imperative to have a clear and concise idea of the objective or goal.

1.3.1.1 Objective Characteristics

Clearly there could be many types of objectives according to the nature of the project, however, it is more usual to have various objectives for a certain set of projects. In any case, all objectives must be taken into account and the problem then becomes a multiobjective decision problem. For instance, there could be a set of alternatives to manufacture a product in accordance with a specific annual demand (first objective, maximize production), while aiming at a minimum cost (second objective, minimize manufacturing costs, especially work force), as well as keeping inventory of product components at the lowest levels possible (third objective, minimize stock of inputs to production), and with a higher quality (fourth objective, maximize quality).

Sometimes these objectives are contradictory among themselves, as it would be for instance to develop an urban transportation system with high efficiency and with frequent service (first objective, maximize quality of service). However, at the same time, with the goal of decreasing the number of buses on the streets to diminish the traffic congestion produced by buses and trams (second objective, minimize the number of buses on streets). That is, a complex situation that was very well defined by a journalist when he said that we wish "*to have urbanism free of obstacles, but at the same time a very efficient transportation system*,"[5] or translated in more colloquial terms it "*to make an omelette without breaking the eggs*".

Consequently, to get something it is usually necessary to give up a bit at the same time, and this is the essence of decision-making. It is almost impossible to get everything we want if we are not able to give up on something in return, therefore, there is some sort of balance and the trick is trying to obtain the best balance.

In general, be it a maximization or a minimization objective, the main purpose is to look for the optimization of that goal. It is not enough to state '*minimize costs*' because we have to indicate the composition of costs for that objective. For instance, if the project analyzes different types of product manufacturing, it is necessary to specify how much it would cost to produce each one.

[5] Las Provincias, Valencia, Spain, June 06, 2008.

1.3.2 Criteria and Thresholds

However, there is more. If for instance the objective calls for the minimization of damages to the environment produced by the construction of a thermoelectric energy plant, that is fine, but this is only an expression of what is wished; it is a very imprecise or blurred concept, due to lack of precision.

Consequently, once the objective is fixed, whatever it might be, it is now necessary to establish conditions to reach that objective. Something else is needed, that is, there must be a constraint establishing, for instance, that there will be a maximum of NOx contamination permitted, or whatever other constraint, because there are never unlimited funds, or unlimited manpower or unlimited capacity to absorb damage, and most of the time it is not possible to go ahead with disregard of the environment or society. These restrictions are '*criteria*' and there are no projects without them. In a project, they are usually different, varied and sometimes very difficult to identify. A criterion for instance can call for maximization of the Internal Rate of Return, and another for the minimization of impact on the environment, and a third calling for the largest creation of jobs, and all of them for the same project or alternative.

Once these criteria are established they are the parameters used for the alternatives to be selected. All multicriteria decision-making methods use criteria, but in LP there is a difference. All criteria must be restrictive with respect to certain values, which are called '*thresholds*'. For instance, if we are dealing with projects or alternatives that will produce harmful emissions there could be a lower, lesser or inferior limit for that emission. Wait a minute! An inferior limit for something that is harmful? What for, since the ideal will be zero emissions?

True, but not realistic, since every human action involves some damage to the environment, and consequently it is not reasonable to predict zero emissions, when it is known to be impossible in most cases. Naturally, if in the thermoelectric generation plant the discharged gases get somehow stored as in an underground reservoir, there would be zero emissions, but this is an unproven technology, and its efficiency is still under scrutiny.

This is clear, but there is also another reason to put that lesser limit. It can be useful in the future for sensitivity analysis, when the DM could request the information about what would happen if he/she pursues cleaner air and consequently the lesser limit will be lowered a little more. From here, it is deduced that it is necessary to work with actual and approachable projects, not with dubious and unreachable ones. Here the principle is also applied that one must start with a very well defined initial state of the problem to also end with a well-defined state (González Cruz 2008).

1.3.3 Alternative Scores

Now, let us look at another issue. Assume that a city has a social program to provide affordable housing to low-income people, and the budget allows for 5,000 m^2 of floor space. After a survey, there is information about how many families need houses with say two bedrooms, three bedrooms and four bedrooms, therefore, there are three alternatives according to house size and floor space.

One criterion could be for instance that there must be a minimum of 3,500 m^2 of floor space, due to economies of scale, and another criterion with a maximum threshold of 5,000, because that is the maximum allowed as per budget. Concretely, there must be a value attached to each alternative for each of these two criteria, specifying how many m^2 per house contemplates each alternative. This value is the '*score*'. The scores thus defined, could be applied in this example to another criterion such as density of people/houses, or to another criterion such as water consumption per house, etc., or to another such as sewage generated by a household, or power consumption, etc.

1.3.4 Modeling an Actual Scenario

We have now all the necessary elements to approximately mimic reality in a LP model, through a mathematical scheme representing as faithfully as possible the actual problem or situation. This model is defined by a set of equations, which correspond to criteria, by the objective, which must be fulfilled by the criteria, and by the alternatives which scores indicate how well each alternative satisfies each criterion.

This scheme has additional advantages because it obliges the analyst to have an intimate knowledge of the project and to analyse where it can fail, see something that was not seen at the beginning or eliminate a concept that is already addressed in another issue and consequently redundant.

1.3.5 Gathering Information

The decision-making process starts by defining the different objectives, the various alternatives that are often present in a given scenario either in one or in different projects, and criteria. These alternatives must be extensively examined, as well as impacts in both ways, that is individual or in series. The purpose of this step is to furnish the DM with information as complete as possible, not necessarily from the technical point of view, but considering effects and impacts into the social, economic and environmental fields.

1.3.6 Analyzing Data and Information

It is necessary to analyze, evaluate and complete this information if not sufficient. Naturally, the purpose of this analysis is to make sure that every aspect has been covered and considered, spoken, discussed and commented within the team.

Sometimes it is also convenient to have some sort of brainstorming to bring new concerns to the discussion, to avoid as much as possible surprises during execution of the process. The actual examples cited below regarding different aspects of the process document the fact that what probably was naïve to think could happen, in reality happened.

1.3.6.1 Considering Characteristics of Different Projects or Alternatives

At this stage, it is convenient to formulate some questions, such as:

- Is the objective realistic and feasible according to the present day knowledge of the matter? For instance, is it possible with actual technical knowledge to build a fusion reactor? As an example, it is perhaps worth recalling the project that led to the 1989 fiasco when two American scientists declared that they had produced cold fusion in a tabletop experiment.
- What are the probabilities of completing the project within the stipulated time? Unfortunately statistics show that a very high percentage of projects are finished with considerable delays and exceeding budget. A very well known example is the construction of the Channel Tunnel between France and England, which doubled its projected cost.
- Does the promoter have the financial and technical capacity to finance or manage it? Many projects have collapsed because of the promoter's bankruptcy. The skeletons of half-finished buildings in many places bear witness of this fact.
- Now an important question arises: How reliable is the information gathered about sales, demand, prices, damages to the environment, damages to a community, etc.? A very large project in Canada was shutdown long after construction initiation because apparently the rights and claims of native people living in the area and opposing the project were not seriously considered; therefore, the objective was not realistic.
- What are the main unknowns of the project? For instance, is the geology of the area where a pipeline will be built thoroughly known? As a best practice, it is worth mentioning the work previously done in the water pipeline in Denizli City, Turkey. In this analysis eight scenario earthquakes with four different earthquake magnitudes between M6 and M7 caused by two different fault ruptures were investigated (Toprak and Taskin 2007).

1.3.6.2 Defining Impacts, Vulnerability and Degradation They Produce

What is an impact?
What is the meaning of vulnerability? How do we measure it?
What is the meaning of degradation? How do we measure it?

1.3.6.3 Identifying Impacts that can be Produced by Potential Projects or Alternatives

- Have impacts been identified, evaluated either in isolation or in series (that is impacts that produce other impacts, which in turn produce others, and so on), as for instance the negative impact of building a tailings pond, which in turn may have other impacts? A cogent example is the Essequibo Disaster in 1995 in Guyana, when cyanide-dashed effluent broke free from a tailings dam and contaminated the Omai and Essequibo rivers, with an unknown second impact on the native population living in the jungle along the rivers.
- How will the project affect the lives of people living at present in the area? For example, two large hydroelectric projects, the Aswan Dam in Egypt and the Three Gorges in China. The first required a worldwide effort to save monuments, and the second forced a massive relocation of people.
- What are the political consequences of some impacts? For instance, how does it reflect in the public opinion that a compromise made known by a politician during a pre-election campaign is not honoured?
- Remediation measures are also taken into account as well as risks inherent to the project, and there could be the need to incorporate safeguards in the form of actions and mechanisms. For instance if in building a dam there is concern about landslides from the mountains, safeguards have to be considered (as a sad example of how this failed to be done is the Vajont Dam disaster[6]). Another example is the Bhopal tragedy,[7] where no provisions existed for safeguards against accidental or improper operation in a chemical plant.
- Public opinion is a very important issue and the corresponding criteria must be established. For instance, because of not paying due attention to resistance by the population,[8] a Canadian gold mining company in Argentina had to stop operations.

[6] Construction of the Vajont Dam in Northern Italy, the biggest in Europe, was finished in 1963. When filling the reservoir, a lack of stability in the mountains enclosing the lake that had formed behind the dam materialized with the fall of a large quantity of boulders into the reservoir. The water wave produced by this plunge killed thousands of people upwards and downwards of the dam.

[7] The Bhopal disaster in India in 1984 was a consequence of unforeseen circumstances that probably nobody had anticipated. In this case the combination of several factors such as the accidental contact of water with methylisocyanate caused a chemical reaction which, combined with other chemicals, generated gases that could not be contained and escaped to the surrounding area, a working-class neighborhood. It is assumed that more than 3,000 people died and perhaps another 500,000 suffered severe injuries.

[8] A multinational firm intended to mine a gold-bearing local mineral by using dangerous chemical compounds to obtain the precious metal. There was the danger of groundwater contamination due to the tailings that would result from the mining process. The project was halted because of protests by the people in the nearby city of Esquel.

1.4 Supporting the Decision Maker

1.4.1 What are the Criteria and Attributes That will be Used?

The criterion concept has already been defined; however, nothing has been said about it. This is one of the most critical and difficult aspects in any method because it entails determining which viewpoint will be employed to evaluate each alternative. In SIMUS, alternatives are in columns while criteria are in rows. Scores for each alternative can be in cardinal or in ordinal values; in the latter, through a subjective appreciation as for instance: 'high', 'medium', 'poor', etc., and converted into cardinal values using a rule. An important concept to consider is that in selecting criteria, these must represent the effects of the project on diverse sectors of the population and not only those of the DM or the agency promoting the project.

In most cases criteria are subordinate to the goal/s. That is, in a project with different alternatives for an airport expansion for instance, and looking to maximize the commercial operation benefits, criteria surely will be related with funding, minimum quantity of passengers needed for commercial feasibility, space needed to accommodate passengers, size and location of the baggage claim area, etc. However, there could be many more criteria related with other impacts, even if these criteria do not relate with the profit objective, for instance:

- Noise generated by different alternatives. It can be related with the height and orientation of airways for commercial and military aircrafts, and in turn it could depend on the characteristics of each alternative. A good best practice is shown in the analysis made to decide on different expansion alternatives for the Maastricht airport in the Netherlands (Vreeker et al. 2001)
- How does lengthening of the runways to existing forests in the area affect the area?
- Are there enough connecting roads to the airport and with enough capacity to handle more traffic?
- How will the microclimate be affected by the increase of flights and especially because of the concentration of flights at peak hours?

1.4.2 Establishing the Thresholds

Remember that these are metrics to create lower or upper limits to criteria. Normally there are thresholds for:

- Maximum level of air contaminants in mg/l, ppm, ppb, etc. allowed in an area (for instance specifying that the maximum permissible level of NOx is 500 ppb).
- Minimum number of houses (say 278 houses of three bedrooms and 341 of two bedrooms, according to potential purchasers as per market studies), to be built in a housing development (to make the operation profitable from the point of view of economies of scale).

- Minimum water consumption per capita (to satisfy basic needs) in a city (70 l/day-person), as well as maximum (250 l/day-person) (to avoid water squandering).These two measures may correspond to two equal criteria and thus the model establishes the gap of consumption which must be within these limits.
- Maximum number of students in a school room (say 40 students).

1.5 Processing Data

Usually the amount of data collected in large projects and even in smaller is daunting; normally there is a lot of crisscrossed information, interrelations, uncertainties, dependencies, indexes, etc, which is impossible for a person to manage efficiently and to say the least, to extract any conclusion from. For instance assume three projects (X, Y, and Z) and say six criteria (A, B, C, D, E, and F). Project X may have an excellent score regarding criteria A and F but poor performances in the balance of criteria. Project Y can have a good performance in criterion A, better scores in B and D and no performance at all in criterion C.[9] Finally, project Z can have poor performances in criteria A, D and E, better than other projects in C and be equal to project X in C.

Now, if the DM has to make a selection of alternatives based on these criteria, what could he/she make out of this information? How can he/she prepare also a ranking of alternatives, which is often required? In this case, this ranking would be relatively simple because there are only 3! possible permutations. However, think in a 7 alternatives project; there will be 7! = 5,040 possible combinations or rankings! On top of that, it is for sure that the DM will want to explore different scenarios in varying some values such as thresholds. When this picture relates with a complex problem like the selection of environmental indicators at all levels, where there could be literally hundreds of alternatives, solving the problem 'by hand' is utopian.

That is what computer decision models are for. They organize the information and provide answers according to certain methodologies and rules, which are normally different. There are easily more than a dozen different computer models for decision-making, customarily grouped in categories such as the Utility Theory, Outranking models, Preference models and Distance to best solution models, as well as some models based on Linear Programming.

As seen, there are several methodologies to apply. Which is the best? Hard to say. Each one possesses its own characteristics and in reality some are more adequate to

---

[9] This is not strange since not all alternatives, even pertaining to the same project, necessarily must comply with all criteria. For instance, imagine a textile project manufacturing wool and cotton yarn. It is evident that a criterion that specifies the percentage of cotton in a yarn has no application in the wool yarn.

certain situations than others. In the opinion of this writer one of the common main problems encountered in using these different methods or models is the high degree of subjectivity, and from this point of view there is a large advantage in using LP where the subjectivity is by far much less than in other systems and, in some cases, nonexistent. Why the advantage?

Because most methods make extensive use of subjective and sometimes individual and preferential values – and even assuming no biased opinions from the DM and his/her staff – they are only estimates, judgments, assessments, and views, that can lead to false results, can get ambiguous answers and are able to produce wrong decisions.

These methodologies rely on what a person or a group think, say, or command (not ruling out vested interests….), and of course there is no guarantee that another DM or group will get the same estimate. LP normally avoids this problem, especially in weighting criteria, or in assuming that one criterion is more significant than another, because these assumptions are not needed, and results are a logical consequence of data imputed and without a subjective bias. Consequently, it is believed that new approaches must be developed to have results not conditioned to personal estimates, and from that point of view, it is thought that LP is unbeatable. Naturally, no matter the method used – and LP is no exception – there is always some degree of subjectivity, as for instance in the selection of criteria, but the use of the latter method reduces this uncertainty to a minimum.

1.6 Modelling

1.6.1 Selecting the Model to Use

This book develops some concepts and provides some information about the different methods, and how they work in order to let the DM know about the different options, as well as a brief analysis of pros and cons for each one.

The examination covers the most popular methods such as MAUT, ELECTRE, PROMETHEE, TOPSIS, AHP and SIMUS. Probably the reader is wondering what the function of these models is if the final decision is taken by the DM. They work by screening the different proposed alternatives and producing the solution that corresponds to the dominant ones, that is the alternatives that perform better than others in accordance with the criteria established. The difference between the methods lies in the ways they use them for that purpose.

Examples illustrate each method and in some cases the solution reached by the respective software is also shown. The whole idea is that the DM, independently if he/she is an expert in this subject – and usually that is not the case – has all the necessary elements to make a sound judgment based on the advantages and disadvantages of each program and fundamentally a clear idea of how they can help solve the problem.

1.7 How to Approach a Problem

The established best approach consists in building a model (no matter which it may be), but able to reproduce as much as possible the existent situation under analysis. This is done in all cases by constructing the '*Decision Matrix*'. Needless to say, the model will never be an exact replication of the actual thing but no efforts must be avoided in trying to get this likeness. For instance, in the past most projects were only interested in the economic aspect without taking into consideration the social and environmental issues. However, this is essential; because these components always exist in any project, for projects usually affect human life and the environment in different manners and scales, therefore this actual condition must be taken into account, in lieu of considering only the economic or financial result.

It is worth remarking that these methodologies are only tools that furnish valuable information and providing in most cases a set of solutions from where the DM is able to make a decision, i.e. nobody can seriously consider blindly following what a computer model says. Consequently, there is no doubt that the DM is the person who, with the knowledge, information, and solutions provided by a model, must make a decision, helped by the ability that most models have in analyzing various scenarios, circumstances and situations, as well as changing conditions.

Since LP is proposed in this book as the preferable tool, this technique is explained in greater detail; however, this is not a mathematical book and for that reason only the essentials of the method are explained. The reader doesn't need to possess a background in this discipline, for these are not theoretical developments but a minimal explanation of the method for the DM, advisors and analysers to understand. A graphic example is posted in a very simple exercise especially designed for the reader who is not familiar with this technique, to get acquainted with it and to fix some mathematical concepts which are useful to understand the model.

The DM who is aware of the different models and selects the one which is more akin to his/her wishes and needs, will be able to instruct analysts and advisors about using the chosen method. With that purpose in mind, the main characteristics of each model are analyzed, without entering into explanations or mathematical developments.

1.8 Delivering Information to the Decision Maker – Its Analysis, Discussion of Results, Feedback and Final Decision

When the computer model has been selected and run, the results are submitted to the DM who, helped by his/her staff, analyzes them. More often than not there will be many questions on his/her part and, frequently guided by intuition and experience, may agree with the results. However, most probably additional information of the type "*What happens if…?*" will be formulated, and the analyst (and the model) must be prepared to answer them; this is usually known as '*Sensitivity analysis*', and permits to find out how the solution changes (how sensible it is), regarding the variation of some parameters. It could be that small variations of say, price in a project, change the selection that had been made, or the opposite that is that the solution previously found is stable for a certain range of variations, which normally is a valuable asset.

The model is then inputted with new values for this analysis, or perhaps new criteria added and some deleted, or different thresholds, and run again; that is the feedback process typical of systems. To properly analyze sensitivity is one of the features that should be used in selecting a computer model to be sure that possesses this quality. If the model provides a rather poor procedure for a sensitivity analysis, its value is relative.

Last, but not least, when the decision has been made, make sure to keep a documented record of everything, that is:

- Grounds on which the final selection was made,
- Base on which alternatives were considered for selection (for instance in an oil pipeline project):
  - Alternative A: Shortest route, but at a higher cost due to mountain crossing,
  - Alternative B: Largest route, but at a lesser cost because of going through a river valley,
  - Alternative C: Least risk of sabotage,
  - Alternative D: Route that avoids the crossing of environmentally protected areas.

  Document the main reasons for considering these alternatives and keep records of everything and especially the basis, grounds or rational explanation backing up each selected alternative. You could need this information in the future!
- Criteria and arguments used to select them. For instance in the case of sabotage explain why and where this could happen and on what grounds the risk is based, that is, what information suggested a measure of vulnerability?
- Details of the thresholds used for each criterion, and explanation of their origin. For instance in a criterion dealing with water contamination, validate the maximum value allowed by local or international standards to which the corresponding threshold is related.
- Justify why a particular computer software model has been selected. If for instance PROMETHEE has been chosen, explain the rationale behind the values established for thresholds ‘p’ and ‘q’ as well as why a certain transfer function has been selected for each criterion (See explanation in Sect. 3.2.3).
- Document requests made by the DM which must be in writing.
- A closing summary should show the final selection made as well as the ordering and reasons for the selection within this ranking, since it could very well be that the first alternative in the ranking is not selected, and instead the third one has been chosen, but of course reasons should be given.

Why is all this documentation necessary? For several reasons, one of them is to make clear the decision of the DM. This way if in the future there are inquiries about the project, there is documentation to support the decision, or to prove that it was erroneous. Second, as a source of valuable information for future projects, even if they are of different nature. Third, because normally, once a project is finished it is hard to say if another project or alternative could have been a better choice. Consequently, the best procedure is to try to formulate each project as well as possible, and from this point of view reliability of information on how and why a selection has been made play a paramount role.

1.9 Operative Sequence for Decision-Making

To wrap up these comments we believe that it is worthwhile elaborating a sequence of the decision-making process. This does not mean that the process follows a sequential order of activities or actions. There is no such thing, and the sequence shown here is only an attempt to organize the procedure, but is certain to be full of loops, feedbacks and even going back on some preliminary decisions and conclusions. The example in Sect. 5.1 illustrates this last concept.

The 'sequence' is exposed as follows:

1. Establish the *objective or objectives* of the project and specify if individually they call for maximization or minimization.
2. Enumerate and define the *projects, alternatives, plans and programs* that will be scrutinized.
3. With the objectives in mind, think about the *criteria* that will be used to evaluate the alternatives, and establish if individually they call for maximization or minimization.
4. Determine the *score of each alternative to each criterion*. Remember that is common that an alternative doesn't have any contribution to certain criteria.
5. According to (3) establish *limits* or thresholds as lower limit or upper limit.
6. Build the *decision table* in Excel.
7. *Solve* the problem using the Solver ad-in.
8. *Submit the results to the DM and feed back the decision table according to his/her opinion*. Remember that alternatives can be added or deleted, same for criteria, and that scores and thresholds can be changed. Run the model again to see the new results, and have in mind that runs can be made for each change or change all the parameters.
9. Perform a *sensitivity analysis* in anticipation of the questions that most probably the DM will make.
10. Keep a *copy* of the different runs, properly labeled with indication of why the changes were made, and archive them. Relate them with whatever documents the DM issues.

1.10 Conclusion for This Chapter

This chapter is devoted to outlining the strategy which is the main subject of this book, and in so doing it gives a glimpse of the decision-making process especially defining what has to be done, identifying alternatives, impacts and criteria. It also deals with complexity and the responsibility of the decision maker, and hints why Linear Programming is selected by solving the decision-making problem. A great deal of comments pertains to the objectives of the project or alternatives as well as the necessity of modeling real life problems and gathering of information.

1.11 Tentative Listing of Activities Where Decision-Making is Needed

It is an impossible task to list all the situations requiring decision-making; as a matter of fact humans do it continuously. Table 1.1 is a listing, admittedly incomplete, of the different areas where normally decision-making is necessary. It has been compiled utilizing data gathered from the Internet, and as much as possible each application mentions the name of the author or authors and the sources, the journal where the case has been published or the book where it was extracted. Also the methodology used in each case has been added. It is believed that this table can help the reader to find how a similar application can be made to a problem on which he/she is working and perhaps to get some useful ideas. These are mostly actual cases and provide a wealth of information.

Table 1.1 gives about 29 different areas of application with some works done by researchers.

Table 1.1 Application examples of decision-making in different areas

Area	Works
Agriculture	Title: ***Three projects to prioritize*** Author: University of Iowa
Airports	Title: ***A note on the selected multicriteria decision-making. Methods and their applications*** Graduate School of Business, University of Pittsburgh
Civil works: Building construction and public works	Title: ***Gestion de développent des applications de l'information en génie*** Author: La Commune de Meyrin Reference: <http://ecolu-info.unige.ch/recherche/COST/Rapport_COST_Meyrin_97-98.pdf>
Education	Title: ***Aplicação conjunta de métodos de apoio multicriterio*** Author: Monteiro Gomes Revista argentina de I.O. año XI No. 23, page 69–86 – Nov. 2002

(continued)

Table 1.1 (continued)

Area	Works
Electrical distribution	Title: ***Multi-Criteria Planning of Local Energy with Multiple Energy Carriers*** Author: Espen Løken Norwegian University of Science and Technology Title: ***Use of geographic information systems in an environmental impact assessment of an overhead power line*** Authors: Warner, L.L. Diab, R.D Reference: *Impact Assessment and Project Appraisal.* ***20,*** *39–47 - Beech Tree Publishing, Surrey, U.K. 2002*
Electric energy: Hydro	Title: ***Medium hydro-power study project (MHSP)*** Nepal Electricity Authority with technical cooperation from the Canadian International, Water and Energy Consultants Reference: <http://www.south-asia.com/mhsp/mhsp.htm>
Electrical generation: Renewable sources	Title: ***Pre-feasibility MCDM tools to aid communities in priorizing viable renewable energy sources*** Authors: Khaled Nigim Nolberto Munier John Green Renewable Energy 29 (2004) 1775–1791 Title: ***An integrated multi-criteria system to assess sustainable energy options: An application of the Prométhée method*** Author: Fausto Cavallaro Reference: <http://www.feem.it/NR/rdonlyres/9E940C73-A958-4D9C-B5CC-48E940D89829/1448/2205.pdf>
Electrical generation: Conventional	Title: ***Ranking Projects using the ELECTRE Method*** Authors: John Buchanan Phil Sheppard Reference: <http://www.orsnz.org.nz/conf33/papers/p58.pdf>

(continued)

Table 1.1 (continued)

Area	Works
Environment: Indicators Impacts contamination Remediation	Title: ***Multicriteria Analysis for Evaluation of Recycling Strategies in Malaysia*** Authors: Santha Chenayah Eiji Takeda Reference: <http://ideas.repec.org/p/osk/wpaper/0501.html>
	Title: ***Multi-Criteria Decision Support for Integrated Technique Assessment*** Authors: Martin Treitz Hannes Schollenberger Benjamin Schrader Jutta Geldermann Otto Rentz RadTech Europe 2005 Conference & Exhibition Reference: <http://www.radtecheurope.com/files_content/march%20 2005%20papers/treitzpapermarch2006.pdf>
	Title: ***Instrumentos económicos para la gestión ambiental: decisiones monocriteriales versus decisiones multicriteriales*** Authors: Fander Falconi Rafael Burbano Revista Iberoamericana de Economía Ecológica Vol. 1: 11–20 http://www.redibec.org/archivos/revista/articulo2.pdf
	Title: ***Measuring Sustainability: A Multi-Criterion Framework*** Author: Giuseppe Munda Reference: <http://www.dse.ec.unipi.it/persone/docenti/luzzati/italiano/ didattica/measuringSD_munda.PDF>
Environment	Title: ***An integrating decision analysis an inexact mixed integer linear programming approach for solid waste management*** Authors: S. Cheng C.W. Chang G.H. Huang Artificial Intelligence 16 I (2003) 543–554

(continued)

Table 1.1 (continued)

Area	Works
Harbours	Title: ***Multicriteria decision-making using ELECTRE*** *Authors:* Wen-Chih Huang Chien-Hua Chen Reference: <http://www.easts.info/on-line/proceedings_05/2237.pdf>
Highways	Title: ***Analytic Network Process Model for Highway Corridor*** Author: Mongkut Piantanakulchai Reference: <http://129.3.20.41/eps/urb/papers/0509/0509021.pdf>
Housing development projects	Title: ***SIMUS aplicado a la construcción de subdivisiones habitacionales urbanas*** Author: Nolberto Munier Seminario Internacional sobre Rehabilitación de Barrios, Toluca, México, Set. 2000
Industrial location: Factories, offices	Title: ***An Application of the Analytical Hierarchy Process to International Location Decision-Making*** Authors: Walailak Atthirawong Bart MacCarthy Reference: <http://www-mmd.eng.cam.ac.uk/cim/imnet/papers2002/Atthirawong.pdf>
Merchandise distribution	Title: ***The MCDM Based Redesign of the Distribution System*** Authors: Jack Zak Hanna Wlodarczak Marcin Kicinski Reference: <http://www.iasi.cnr.it/ewgt/13conference/119_zak.pdf>
Military applications	Title: ***Una aplicación del ELECTRE a la selección de un caza-bombardero*** Author: Carlos Romero (1996) – See Bibliography

(continued)

Table 1.1 (continued)

Area	Works
Mining	Title: ***Multicriteria choice of ore transport system to an underground mine: application of the Prométhée method*** Authors: B. Elevli A. Demirci Reference: <http://www.saimm.co.za/publications/downloads/v104n05p251.pdf>
Pipelines and aqueducts	Title: ***Water supply system decision-making using multicriteria analysis*** Authors: Danielle C Morais Adiel T. Almeida Reference : <http://www.wrc.org.za/downloads/watersa/2006/April%2006/1869.pdf> Title: ***Baku-Tbilisi-Ceyhan pipeline*** Author: Mark Tran <http://www.guardian.co.uk/business/2005/may/26/businessqandas.oilandpetrol>
River water use: Dams, irrigation, distribution	Title: ***Colorado River*** Author: Massachusetts Institute of Technology
Selection in general	Title: ***Strict uncertainty: A criterion for moderately pessimistic decision makers*** Author: Enrique Ballestero (2002) Decision Sciences Winter 2002; 33,1 ABI/INFORM Global
Services evaluation	Title: ***DTLR multi-criteria analysis manual*** Reference: <http://www.sfu.ca/mpp/pdf_news/811-04%20UK%20MCA%20Manual.pdf> Title: ***Multicriteria Approach to Decision Aid: Prométhée & Gaia*** Author: Bertrand Mareschal Reference: <http://theses.ulb.ac.be/ETD-db/collection/available/ULBetd-06162008-172523/unrestricted/Annexe_PROMETHEE.pdf>

(continued)

Table 1.1 (continued)

Area	Works
Social projects	Title: ***Community infrastructure upgrading in Ghana*** Author: Nolberto Munier (2004)
Stocks portfolio	Title: ***Selection of a portfolio management*** Author: Unknown
Tunnels, bridges, viaducts	Title: ***Scheduling for bridges repairs*** Author: Nolberto Munier
Urban basic infrastructure: Sewers, water treatment plants, pavements	Title: ***Scheduling for municipal works*** Author: Nolberto Munier
Urban highways	Title: ***Selection of alternatives for an urban highway improvement*** Author: Nolberto Munier
Urban land use	Title: ***Multicriteria Evaluation of Sustainable Agricultural Land Use. A case study of Lesvos*** Authors: Cerda Hermanides Peter Nijkamp Reference: <http://ideas.repec.org/p/dgr/vuarem/1997-5.html>
Urban rehabilitation	Title: ***Rehabilitation in the city of Leon, Spain*** Author: Nolberto Munier
Waste management: Collection, landfills, incinerators	Title: ***An integrated multi-criteria decision analysis and inexact mixed integer linear programming approach for solid waste management*** Title: ***Selection of location for two wastes incinerators*** Authors: Laura Tasca University of Milano, Italy

References

Ballestero, E. (2002). Strict uncertainty: A criterion for moderately pessimistic decision makers. *Decision Sciences, 33*(1, Winter), 87–107.

Dantzig, G. (1948). *Linear programming and extensions*. Princeton University Press, Princeton, New Jersey.

Gonzalez Cruz, M. (2008). *Teoría y Metodología del Proyecto – El proyecto como proceso de resolución de problemas*. Universidad Politécnica de Valencia, Departamento de Dirección y Gestión de Proyectos, España. Retrieved September 8, 2009, from website: http://www.upv.es/index-en.html.

Munier, N. (2004). *Multicriteria environmental assessment: A practical guide*. Dordrecht: Kluwer.

Romero, C. (1996). *Análisis de las decisiones multicriterio*. Madrid: Isdefe.

Toprak, S., & Taskin, F. (2007). Estimation of earthquake damage to buried pipelines caused by ground shaking. *Natural Hazards, 40*(1), 1–24.

Vreeker, R., Nijkamp, P., Ter Welle, C. (2001). Discussion paper: A multicriteria decision support methodology for evaluating airport expansion plans. *Tinbergen Institute Discussion Paper TI2001-005/3*.

Chapter 2
Getting and Processing Data

Abstract The different intervening elements for the decision-making process are commented on here. Its only purpose is to show the whole scenario of factors involved, a description of each one and their ordering.

Keywords People • Stakeholders • Competing projects • Criteria weights • Threshold standards • Impacts

2.1 Data Collection: Background Information

Since normally in the decision-making process, uncertainty exists and there are subjective aspects to consider, the methodology should be a doctrine that works following these principles:

- Exert the greatest possible objectivity, which involves reducing personal perception at a minimum coming from the DM, the decision-making entity or analysts. These assessments involve estimating weights of criteria, and in some models expressing preferences of one action over another, or fixing some indifference or acceptance levels. Because these are subjective perceptions, even if based on a sound reasoning, results may change when values from one DM are confronted with results from another, and even perhaps including the original DM when approaching the same problem some time later.
- In the same manner as engineering projects develop in a way to enable somebody in the future to go over the original calculations in order to check if these were or were not correct, it is understood that in decision-making practice, even with subjective aspects, there must be a document supporting the reasons for taking any decision. That is, it is not enough that somebody could think that A is better than B, because it is necessary to have a coherent and reasonable justification about the reasons to articulate this statement.

N. Munier, *A Strategy for Using Multicriteria Analysis in Decision-Making: A Guide for Simple and Complex Environmental Projects*, DOI 10.1007/978-94-007-1512-7_2,

Whatever the method used it must strive to obtain transparency of the whole process in such a way as to dissipate any doubt about what was done and on what basis. Naturally, even with the best goodwill, knowledge and expertise, a mistaken decision can be made, but if it is known on what grounds that decision was adopted, at least there would be some possibility of finding an error and either correcting it, or avoiding it in the future. Otherwise it could be impossible to further verify if something went wrong, either because the person who made the decision is no longer around, or doesn't remember (perhaps deliberately) the reasons that led to a certain decision. If decisions are documented, it is easier to reveal causes in the future. The Challenger disaster[1] shows tangibly how this system works.

- In any project there are values which are completely subjective, such as for instance visual impacts, trouble caused by continuous trembling (for instance by passing trains), or the loss of value of a property as a project impact. However, there are mechanisms that can be adapted to compute approximately those facts, like surveys, polling public opinion, for the first two impacts, and the use of the Delphi[2] method for weight determinations, or the hedonic[3] method for the loss of value of a property, just to name some of them. But even if the decision is based on the DM's personal experience, it is necessary to document the reasons supporting it.

The strategy proposed here tries to accommodate these perceptions and, while trying to avoid subjectivity, we assume that the DM is the one who takes the final decision, not the model that has been chosen. This strategy does not interfere with application of this human attribute to the process, but instead attempts to support the DM's decision in adopting one or another project or solutions method, based on certain and reliable information as well as a transparent procedure.

---

[1] It refers to the Space Shuttle Challenger disaster that killed seven people in 1986. Jenny Jones, in the 'The Madison Courier. Com', 18 Feb. 2005, comments:
"*So Boisjoly* (one of the engineers working in the firm contracted by NASA) *drafted a letter to his managers, stating how faulty the O-ring joint system was and the effect it could have on flights to come. "It is my honest and very real fear that if we do not take action to dedicate a team to solve the problem with the field joint having the number one priority, then we stand in jeopardy of losing a flight along with all the launch pad facility," Boisjoly wrote. "The result would be a catastrophe of the highest order – loss of human life*".

[2] Delphi method: Developed after WW II by the Rand Corporation. It is essentially a system that works with groups of experts, spatially located, in order to obtain independent reports about certain issues and responding to a coordinator.

[3] Hedonic appreciation. The appraisal is based on an estimate of changes in the price of a good in the market such as a house, when these changes affect one of the attributes characterizing the good. For instance an attribute might be the quietness of the place, or a beautiful landscape, or the enjoyment of wildlife. These attributes could be affected, for instance, as a consequence of construction of a highway nearby leading to a decrease in the value of the property.

2.2 People Participating in the Selection Process and Sensitivity Analysis

We point out here that, whatever the method adopted for solving multicriteria problems, four entities participate in this process, and the proposed methodology is no exception. These four entities are:

- The decision entity, that could be either a person or a group of persons who hold, together with stakeholders, the decision power to carry out the selection,
- Technicians who supply quantitative and qualitative information,
- The analyst or group of analysts who process the data and who are knowledgeable about suitable mathematical procedures,
- Citizens.

The three first entities must complement each other, since none of them can make a rational decision without the support of the others. For instance, the decision entity may be eager to execute a series of projects contemplating many factors. However, it could be taking the risk that the outcome is not the expected one because does not consider technical issues, nor know if the plan is one of the best or even feasible. Technicians are not able to make a decision because they do not have a bird's eye view or a complete scenario involving the project, and do not know the compromises assumed by high-level officers. Besides, they usually do not know the methodologies or models to apply. The analysts can perform an impeccable job, but without information from the DM and the data by technicians, it is possible that the result would be poor.

On the other hand, it could appear to be a nice, elegant and reasonable result or conclusion; however, it is also necessary to perform sensitivity analysis to make sure that the solution agreed upon holds stable when some of the parameters change. These changes could come from the DM or from other sources, as for instance a variation in market conditions. That is, it is evident that even if the solution reached conforms to everyone's requirements, it cannot be adopted without analysing its sensitivity because it could very well be that a change in variables such as demand, price, funds availability, etc, which in most cases may vary between large limits, invalidates the solution reached. In that case it is logical to start looking again because the solution doesn't pass this test.

As an example, a DM could need information about the relative importance of each criterion, because with that knowledge it could be convenient to increase or diminish the weight of some of them to better meet a certain purpose. As a bottom line, once a compromise solution has been agreed upon it is mandatory to carry out a sensitivity analysis that will allow all participants to know the answers to questions of the type '*What happens if…?*'

Another fundamental piece of information for the DM is to learn how stable the solution is. If the solution found is very sensitive to changes in the restrictions, or in the addition or deletion of alternatives, or in the values of the scores in the decision matrix, probably it is not convenient. The reason being that most cases do not develop as planned or in the way one thinks that they will behave; there are unaccounted for factors that can vary or change, either considered or not, and also there are chances that new ones will appear. This does not mean that a 'good' solution must admit all changes without variation, for that is almost impossible. What it is meant here is the necessity of determining the amount or quantity of variation that the solution can stand without changing substantially.

If a change in assumptions, scores, criteria or thresholds provokes a light variation in the outcome, the solution may still be accepted, but if by augmenting the price or equipment say by 5%, the solution changes, then it is not really stable and perhaps it is better to look for a new one.

There is another benefit in performing sensitivity analysis, because if it is true that the LP algorithm is mathematically correct and will always give the right answer (according to the quality of data), we cannot be so sure about externalities, because changes in demand for instance, could be unpredictable. We enter here into the domain of risk analysis, and it can be advantageously used to provide '*safeguards*' that can drastically diminish the risk, however, this is a topic that does not belong here. We state only that, if a solution is good but offers some degree of instability due to risk, and there appears to be no viable alternative, it may be possible, usually at greater cost, to reduce the risk to acceptable levels.

As an example, a city was considering crossing a 150 m wide river for the extension of a subway line. Two alternatives were studied; one of them was to use a Tunnel Boring Machine that had been used in digging the subway tunnel, for extending it below the river. The second alternative consisted in digging a trench in the bed of the river, which would embrace 15 prefabricated concrete parts to form a tunnel for the subway. Each alternative had its own characteristics, costs and problems. However, the first alternative with a tunnel under the river raised the hazard of water percolating into the tunnel, with a big risk to human lives and equipment. A safeguard was devised and consisted in pouring a concrete slab on the bed of the river. That would eliminate the problem in the first alternative and could perhaps influence the decision process because it was a much cheaper alternative than the second one.

It is then clear that it is important to examine in some detail all components of a complex project and the different circumstances that reasonably can appear. It is all too common to perform an analysis, study statistical series and prepare reports and forecasts, all based on assumptions that are not realistic, or that simply ignore or omit some facts and procedures, such as requesting various people's opinions on a crucial issue. Expert panels are not always complete or truly representative and obviously each member of the panel will express a personal opinion based on acquired experience with related issues. Assuming that the panel has been appointed with some reasonable degree of thought, it is probable that these opinions will be well founded; on the other hand, they may not necessarily reflect the public opinion, and that is understandable because the panel generally does not 'live' the problem, as do the people who will be affected by the project. Look at these real life examples:

In the selection of three alternatives in a road project in a large city, the DM and team adopted one of the alternatives, considering all the aspects that the experts found relevant. Just to be on the safe side with regard to their assumptions, they conducted a survey to learn the point of view of citizens in general regarding the project.

Surprisingly, the criterion related to impact of one of the options in the urban area, showed values completely different from those assumed. The reason? The experts had not considered that said option would divide a densely populated area in two, cutting off direct communication for its inhabitants, which would have made life more complicated for neighbours in their daily travel to work, schools, hospitals, supermarkets, etc. That is, only when the experts consulted the people who would live with the consequences of the project did they realize that they had neglected to consider a vital connectivity aspect.

A similar problem arose in the Spanish city of Valencia in 2009 with the construction for the F1 circuit. The news reported that the Nazaret's[4] Neighbours Association complained, "*The European Grand Prix constitutes an architectonic barrier between the maritime district and the Nazaret area*".[5]

In an Argentine city, City Hall officials studied different options to improve life for people living in shantytowns. They then approved the option that consisted in building new houses for these people on urban land located on the outskirts of the city. They built new modest houses and families from some shantytowns relocated there; surprisingly, and even though they now had decent dwellings, many people complained, including some families who opposed the relocation scheme and stayed in the old site. After a couple of months several people from the new town came back to the old premises, now destroyed, and built new dwellings from cardboard, discarded wood and zinc roofs held in place by big stones…. That is, the old site little by little became the same as it was before. What is the lesson extracted? Simply, the scheme failed.

It failed, because the DM did not consult the people most directly affected, and arbitrarily and politically decided to relocate them. One naturally can be surprised that people returned to less pleasant dwellings, but there were compelling reasons for that behaviour. Without going into detail, this behaviour can be partially explained by a few cogent aspects. For instance: Proximity to sources of income, which albeit often very modest and precarious – such as collecting hardboard from discarded boxes and selling it. Such facts dictated their way of life; logically it was of paramount importance to live near that source of income, since the new location was 7 or 8 km away and meant additional expenses in transportation. Considering other aspects such as long-term relationships, nearness of schools, etc., might help to understand people's reluctance to relocate. Wrong conclusions and misguided work would have been avoided if people had been consulted.

This second procedure was used in the shanty town of Nezahuacoyotl on the outskirts of Mexico City; the DM learned after consultations with local people that they did not want to be relocated and asked instead for help to improve their living. Consequently, an agreement was reached for the government to provide basic infrastructure to the neighbourhood, such as sewage, pavement, electricity, and water, which would be repaid over a generously long period. Needless to say, in the

[4] Nazaret. A neighborhood in the city of Valencia, Spain.

[5] Levante, September 02, 2009.

Argentinean case the DM, by not considering the social aspects of the problem, created a ghetto isolating people from the social fabric of the city, instead of integrating it such as in the Mexican case. Another example can be similarly analyzed by examining case number 52 for several cities in Ghana in a project sponsored by the World Bank.

As a bottom-line conclusion from these examples, all projects, whatever their nature, are built direct or indirectly for people, and so people have to be consulted about them. These cases show that public opinion is fundamental for the realization and success of most projects (Stolp et al. 2002).

Needless to say, the objective or objectives must be clearly stated as mentioned in Sect. 1.3.1. At this stage it is important for the firm,[6] before committing to a project, to make an honest appraisal of its ability to proceed to its conclusion. From this point of view it is useful to perform a SWOT analysis, that is for the firm to determine its technical, financial, economic and expertise capabilities (**S**trength), as well as its vulnerabilities (**W**eakness), for instance in manpower.

The firm must also examine the chosen scenario from beginning to end, considering factors such as government legislation, perhaps recently enacted, that could economically benefit the project (e.g. a potential government subsidy for each new employee in a new industrial plant that is an object of the project); that is, the firm must consider (**O**pportunities). Last but not least, the firm must be aware of a project's hazards, risks or vulnerabilities, for instance the chance that somebody else can simultaneously manufacture a better or cheaper product, or import it, (**T**hreats). There is enough literature on this subject that it need not be explained in detail here, but we mention it to ensure awareness that the firm must know its position and must realistically assess its chances for success or failure.

2.3 Nature of the Competing Projects or Alternatives

We refer here to the main characteristics of a project; their classification is important because criteria to evaluate them are based on these features. From this point of view a list of examples follows, but clearly the universe of projects is larger than this.

Geo-Referenced projects
As examples are those related to the construction of service centers, for instance health centers, distributed within the city's spatial area, where elements such as population density, average distance travelled by the user, number of inhabitants in the area, etc., play a fundamental role. These projects appear in those cases where City Hall must decide on costly undertakings related to public health, education, or

[6] From now on when reference is made to a person, office, company, firm, entrepreneur, consulting office, municipal or government agency, etc., in charge of developing a project, we will use the word '*firm*'. We also assume that the DM belongs to the firm.

sports, i.e., choice of what part of the city, amongst all the available sites, to build hospitals, schools and community centers, in order to maximize the population that will benefit from the service.

In these circumstances the project may be supported by a GIS (Geographic Information System) whose contribution can be invaluable, not only in this case but also in many others. For instance in a rural agricultural project for determining the best rotation of crops, the GIS may contribute with information about the present day situation of crops, the number of hectares cultivated with each one, kind of crop, existence of irrigation ditches, their flows and routes, etc. In urban economy projects, a GIS is normally used for cadastral purposes to identify changes in parcels that have not been declared for tax purposes, and also to determine empty areas that can be used for housing complexes to increase density.

An important factor in the planning of an urban project is to determine the location of buried utilities. Precisely this lack of information usually leads to complicated and long and very often costly procedures to determine competence and responsibility in repairing telephone and energy cables, sewers and old conduits, water mains, etc., which can be accidentally damaged or even destroyed by machinery. In knowing with certainty what lies underground, provisions can be made long before a project starts, in order to communicate this knowledge to other interested parties about the work to be performed. Usually this is reflected in the final cost of the project under consideration, and especially in its completion time. Therefore, good planning that contemplates different issues helps a lot in finishing the project within cost estimates and as scheduled.

Feasibility

The fact that a project is sound and has proved to be economically, socially and environmentally feasible does not necessarily mean that it is valid; there could be factors that were not considered that could make the project unfeasible or more costly. As an example, assume that the project is an urban plan to build a new housing development with 1,050 family houses. The project gets flying colors from every point of view, except that nobody realized that the closest sewer trunk is 2.8 km away; there are then two alternatives:

(a) To contact City Hall to find out if it is possible to connect, at the entrepreneur's expense, the development with the existing sewer trunk, a costly undertaking which also implies breaking down 2.8 km of paved road, laying the piping, covering it and repaving, or
(b) To build a small sewage treatment plant for the neighborhood, which is also quite expensive.

In the first case it is also necessary to find out if the existing sewage trunk has enough capacity to absorb the sewage from the 1,050 houses, amounting to about 1,260 m^3/day, and, of course if the city sewage treatment plant can also accept this additional load. Otherwise, the neighborhood could end up with certain periods during which its streets are flooded with sewage because the main system collapsed.

In the second option, everything is fine, however, where will treated sewage be discharged? Is there a river or creek nearby where this can be accommodated? If not, the entrepreneur is back to the first option. For some people this type of problems seems unlikely to happen in real life, but they do. From these cases it transpires that the DM has to make his/her best efforts to consider every nook of the project itself but also the surroundings, setting or milieu where the project will be immersed.

Sustainability
Their objective is to reach urban sustainability and involve a series of political, economical, social, and environmental issues as well as conservation of natural resources. These projects are common in Europe, especially under the Aalborg Commitments Plan.[7]

Urban rehabilitation
Normally this type of project relates to taking advantage of old and no longer used railway yards, or abandoned maritime or river wharves and service areas (offices, warehouses, roads, etc.). Customarily there are different issues involved in these programs such as spatial and environmental planning, including subjects such as job generation, transportation, communications, recreational areas, parks, etc. More often than not, the city government is very seriously engaged for different reasons. These reasons stem from the desire to recover neglected zones which are typically located in a very central part of the city and dedicating them to business, such as Canary Wharf in London, or to financing such as in La Defense in Paris, hotels and residences such as in Puerto Madero in Buenos Aires, etc. Example number 38 in Table 9.9 refers to a case in a Spanish city, as well as case number 59 for the city of León, México.

Social
Social projects, such as those already commented on above. See also in Table 9.9, case number 55 for community development, as well as project number 56 to decrease pedestrian and cyclist street accidents rate in a city, and also number 58 to select the location of a health center.

Macro
They are mainly related with urban and metropolitan development plans including for instance a Municipal Territorial Ordinance Plan, the institutional transformation

[7] Aalborg's Commitments. In 1994 in the Danish city of Aalborg was born a movement focused on obtaining a consensual declaration from European Cities and Villages towards a Local Urban Sustainability, and following directives from United Nations Agenda 21. In 2007, more than 500 governments had signed the agreement to improve the environment in their respective communities. This fact has had and continues to have profound repercussions not only in the environment but in the social sector too, since no project is given the green light if it is not accompanied by a thorough environmental, social and economic study (that is sustainable), and in accordance with directives from Agenda 21. This in turn, has provoked some heavy activity in projects selection, which is the subject of this book.

of the agency in charge of providing water and sewer services, etc. Within these great lines there can be hundreds of projects that in turn may be subdivided into many others. For instance, within the first mentioned category are projects intended to:

- Recover unused railway tracks and conversion of the vacant space into a linear park,
- Increase safety in an area due to high crime,
- Reversal of contamination of city aquifers,
- Execute an industrial cadastre or housings or empty spaces,
- Establish bus communication between different cities and villages of the metropolitan area.

2.4 Relationships Between Competing Projects, Alternatives or Options

This refers to relations or links that can exist between projects. From this point of view, there are:

Relationships denoting *precedence* and *continuity*.
When there is a dependency between subprojects pertaining to a larger project, for instance construction of a hydroelectric plant. There are possibly two large subprojects: (a) construction of the dam and (b) construction and installation of the power plant, and perhaps a third one that could be ancillary works to divert the river to build the dam.

It is obvious that there is a question of precedence since the power house cannot be built if the dam has not been erected first, and in turn, the dam cannot be built if the river has not been diverted. In this area, in the very complex project of construction of the 14,000 MW Itaipú Hydroelectric Project in Brazil, the whole project involved a series of 'smaller' ones, built in sequence, which probably started with the necessary digging of a temporary channel[8] to reroute the water of the Paraná River, and only after that, and once the enclosure was dried, was it possible to start building the dam, which has a height of a 65-story building.

This case of Itaipú was brought up not to illustrate the decision-making procedure[9]; it was only mentioned to exemplify precedence and continuity in an actual large undertaking. Then, it is necessary to inform the decision model about this precedence because if not, two things can happen (a) the model selects only one of the linked projects, and/or (b) the model indicates an inverse sequence of construction, which of course is not possible.

[8] To bypass the huge Paraná River (the seventh largest river in the world), this trench was a little more than 2 km long and involved removing 50 million tons of earth.

[9] Of course, there are decisions to make when analyzing a complex project, and following with the hydroelectric example, there could be several alternatives, for instance having two or three different ways to dry the place for constructing the dam (like using cofferdams instead of a diversion).

In reality, dependency exists in all projects since, to perform a study of financial feasibility for instance, it is mandatory to execute a series of steps, (or micro-projects) such as calculations for working capital, costs, investments, etc., which can take many days and no little efforts. For instance in a large environmental project in Missouri, U.S.A. for the de-commissioning of an old nuclear-chemical plant, it was necessary to first undertake many technical studies, especially related with radioactivity residues, before cleaning some very contaminated areas and before deciding what to do with the contaminated liquid and solid waste. Naturally, there are very well known and proven techniques with suitable software for planning and scheduling, such as Primavera® (Oracle), and Project® (Microsoft) to deal with this sequencing, but it is only mentioned here to make the reader aware that this sequencing, especially in large projects, has to be considered and included in the model.

Exclusive and compatible projects
This type of project or alternatives appears when a decision has to be taken on incompatible ones. This is the case when alternatives are mutually exclusive as in the case of evaluating either digging a tunnel or building a bridge to cross a mighty river. But also it could be that that is not the case, and that both projects are envisaged and both necessary. Whatever the case, the model must be instructed to consider them as either exclusive or complementary.

Congruent projects
In other circumstances one needs to check another kind of compatibility, this time within the same plan. For instance, a plan to build two urban intersections A and B, separated by ten blocks and on the same avenue. In this case A could be a bridge and B a tunnel or both with the same structure, it really doesn't matter. But what really is important is that both must be congruent from the point of view of vehicle capacity per hour in order to make sure that there will be no possibility of bottlenecks because one of the intersections can handle less traffic than the other. As a minimum they have to be equal, but it could also be the case where one of the intersections, say B, also receives traffic from another avenue, and in this case, logically its capacity should be larger than A. Naturally, any good engineering study will have these concepts covered, but the intention in mentioning them here is to pinpoint that the decision model has to be 'told' about them. There is a related example in project number 54, Table 9.9.

Complementary projects
As an example of complementary projects consider two alternatives to link the continent with an island, these being a bridge or a ferry. May be the two projects are needed, because they are separated by a considerable distance and traffic is very intense. A system like this exists in the city of Vancouver, in Canada, where the suspension bridge and a ferry system are used simultaneously to link the mainland with the North Shore, and each complement the other because the bridge is used for vehicular traffic while the ferry transports passengers.

Package of projects
Sometimes there are cases in which the result can show a selection of projects that must act together, not isolated one from the other. For instance, assume that a city is worried about the number of transit accidents affecting pedestrians. A list of different measures or solutions includes options like establishing courses to educate the citizen, media advertisements, construction of road bumps that force vehicles to decrease speed, more police surveillance, more traffic lights, etc. If this information is fed to the mathematical model, the model can choose amongst the different alternatives, possibly providing a set portfolio of combined measures that can work well together.

Projects financially related to time
In this type of project, considerations must be given to both total duration of projects and their timing. This last observation derives from the fact that many projects exceed in their duration the fiscal year, and consequently, there is need to indicate to the model how much funds are needed and during what period. Otherwise, it is possible that a project is started but not continued in its second year since the decision model cannot extend its execution beyond one time period, and this could happen because it has not been alerted about the continuity in time and funds of this project, to a succeeding time period.

Operation safety in projects
In all projects, besides the technical, economic, environment and social restrictions expressed by criteria, it is mandatory to consider criteria related with safety. Many projects have hazards and safety risks such as in construction of skyscrapers, large bridges, tunnel digging, chemical plants commissioning, etc., and safety measures must be considered for each project or alternative. The well-known Bhopal disaster in 1984 in India comes immediately to mind, because of the accidental release of toxic gases, and also the mishap in Basle, Switzerland in 1986 when a chemical plant discharged a huge quantity of dangerous chemical products into the Rhine River.

There are also projects where a selection has to be made between replacing a facility – that is building a new one – for instance road bridges, or improving an old one to increase its useful life. However, it is necessary to also consider if a 50 years old bridge continues in good condition, especially taking into account not only the higher speed of vehicles, which has an impact on dynamic loads, but also if it is able to safely support the weight of large trucks, completely loaded and with semitrailers. Perhaps there exists an alternative to allow the bridge to continue to be used but only for cars, creating a need to build another one to handle heavy vehicles. Clearly this is a complex decision problem.

Homogeneous and heterogeneous
There are activities that involve making decisions on a large range of projects, which is typical for large organizations such as a City Hall, a Bank, developers, industries, etc.

The number of projects can span from one to hundreds, and can be of the same type or not, for instance in the urban case there could be a homogeneous set of independent projects pertaining to the following areas:

- *Infrastructure*, such as road construction, paving, area rehabilitation, river cleaning, etc.
- *Social*, as for instance plans to get children off the streets, measures to reduce road accidents, etc.
- *Urban strategic planning* such as identifying urban indicators, or to determine municipal policies regarding recycling of residential wastes.

It is also possible for a heterogeneous set to materialize in a blend of all the above mentioned projects, as is normal in a plan of urban development in both short and long terms, where there are budgets established for each municipal department on which the corresponding projects depend. Thus there is a budget for the Health Department, which for instance has a 5 year plan for the construction of three general hospitals, one specializing in cardiology, plus the construction of 25 health centers, as well as vaccination campaigns against the flue, poliomyelitis, etc., and has assigned weights, that is importance levels or priorities to each project, since there is little chance that all projects have the same importance.

Each municipal department may have plans and similar budgets in their own areas, and the five-year plan could include all the projects from all departments, which can add to hundreds as mentioned. It also could be that weight has been assigned to each department, which probably keeps a relationship with budgets assigned to each one.

On the other hand it is necessary to know and detail City Hall's genuine resources, which can also be from heterogeneous sources, such as:

- Coming from real estate taxes, vehicle taxes, charges applied to getting diverse documents and activities like issuing marriage certificates, birth certificates, etc.,
- Funds from provincial, state or community participation and according to a certain criterion, for instance the number of people in each city,
- Funds from the federal government or federal co-participation.

Feasible

Needless to say, proposed projects can be utopist or unreachable. The first case would be for instance a project on human development aiming to create equality in the social sector, related to education, economy, housing, etc., commendable of course, but not realistic. By the same token it is not possible to undertake unreachable projects, such as generation of electric energy through fission in 2 years time, when the technology is still trying to determine how to do it.

Dependency on inexistent infrastructure

Another type of dependency is that which derives from the lack of a service to develop a project. For instance, is there enough manpower in a certain region for a car manufacturing plant, or do workers have to be brought from other sites? Of course, this can be done, but at what cost?

Political or imposed
Certain projects must be included in the final strategy selection, whatever the reasons. If it is true that pre-electoral speeches and promises are very often forgotten or 'postponed', we must also admit that they have to be honored. Consequently a set of municipal projects, including a promised project such as '*Construction of storm sewers to eliminate flooding in the East area of the city*', must mandatorily be in the final agreed portfolio of projects, and this obligation ought to be introduced into the mathematical model.

In conclusion, a question: In case that one or more projects have several options or alternatives, should all of these be analyzed independently? The answer is positive, because one of them could be more convenient than others and then can be chosen in the final result. Another question: Is it possible to assign weight to each alternative, project or option? Yes, it is and is usually done when it is convenient that important alternatives be part of the ultimate solution. For instance, if we have two projects, one of them '*Stop flooding in the South area of the city*', and the other '*Enlarge the sidewalks on 3rd avenue*' it is obvious that the first project deserve more attention and priority than the second and consequently would merit a higher weight.

2.5 Analysis of Competing Projects or Alternatives

There are variations about this subject based on different configurations. For instance there could be:

1. A large project, such as the rehabilitation of a derelict area in a city in a plan comprising offices and residences in high rises, as well as amenities and green spaces. There could be several alternatives as for instance:

   (a) High-rises with a blend of offices and residences,
   (b) Dedicated, lower high-rises for offices and taller high-rises for residential use,
   (c) Alternative (a) with more amenities added than in alternative (b),
   (d) Alternative (b) with more green space than alternative (a).

   The objective is to maximize both, the profit and the use of land.

   It is evident that there is a strong relationship between these different alternatives especially in the use of available resources (money, time, permits, etc.).These inter-relationships between alternatives need to be analyzed because in some cases one may preclude another, or perhaps it is possible to make a blend of alternatives.
2. Municipal plans in a large city usually present a portfolio of completely unrelated plans such us the construction of a hospital and the expansion of the water treatment plant. However, most probably they will share some resources and these must be analyzed.
3. Another case could be the construction of temporary facilities (construction camps) for a very large project that takes place along several kilometres, such as the construction of an aqueduct where its construction goes on simultaneously in three different sections of the total distance, and separated by perhaps hundreds of kilometres. Each subproject may have at the same time

diverse alternatives concerning number of lodging facilities, size of the kitchen, type of sewage to be built, etc.

4. Another type of projects, already mentioned, are exclusive, that is if A is built, B cannot be built; for instance, a project to link two cities by either a highway or a freeway.
5. There could also be manufacturing projects. Assume for instance that a firm manufactures washers and dryers. Regarding washers, they have three different capacities and the same for dryers. In addition, for washers they have two models; either with horizontal or with vertical drums, and within the vertical types those with fixed drum and agitator.

   A market study done for the period, gave an estimated demand for each type and size of washers and dryers. It is most probable that all models shared the same components for instance same quality and thickness of steel sheets, electric motors, belts, gears, hinges, etc., but they have different prices, consume also different quantities of other inputs and yield different profits. The manufacturer may use decision-making procedures to determine quantities to manufacture of each appliance in order to maximize his/her profit.
6. Sometimes there are systems with a large variety of alternatives as is for instance dealing with type of crops and farm products. Even for a criterion in particular, for instance the annual production in millions of Euros, values corresponding to each product may greatly diverge from others. See Table 2.1, which represents the output of a region.

Table 2.1 Agriculture and farm products and annual output values

Products →	Wheat	Beef	Rye	Poultry	Milk	Fish
Criterion						
Annual output in millions of Euros	456.1	270.6	78.9	12.3	185.9	206.8

7. Sometimes, there is a need to establish an interval of values for the different products to compare their outputs. For instance, from 500 to 400 millions, from 400 to 300 millions, from 300 to 200 millions, and so on. Each level is then rated in a scale say from 1 to 5, the higher the level, the larger the rating and these ratings are used for computation. If in the last example it is assumed a cardinal valuation of say 4 for the 200/300 interval and 3 for the 100/200 the interval, it means that the beef output is 1.33 times greater than milk, i.e. 247, which is not true. Naturally the reason for this discrepancy is the assumption of a uniform value for each interval.

   Smaller intervals can be considered but whatever the procedure it is clear that we are working with subjective values. LP, working with the scores assigned individually to each option does not have that problem, which is a considerable advantage especially when this reasoning may be done for all criteria, as is done in some actual cases.

As bottom line it is evident the necessity to perform a rational analysis between interrelationships that might exist between all projects, alternatives, or programs considered. This is important because many times different projects show a technical dependency that should be considered, under the penalty of selecting one of them and not the other. For instance, there can be different municipal projects such as construction of storm water drains in an avenue and then paving it. Naturally, the former has to precede the latter; otherwise, it is possible to end up paving the avenue without doing the underground work. That is, the analysis must reflect a reality and decisions taken accordingly.

2.6 Criteria to Use

Criteria are the conditions or restrictions that projects are subject to, and must be carefully analyzed in order to be reasonably sure that most of them are considered. It is worth noting that the word '*criterion*' is commonly used involving both the attribute and objectives concepts. However, in general, the attribute name is reserved for those criteria or restrictions that are limited by numeric values. For instance, in a restriction that requires keeping an aquifer sustainable, it is usual to establish a threshold or limit value regarding the maximum water flow to be extracted, so it becomes an attribute.

How the DM does settles on which criteria to use? There is not an easy answer to this question, because criteria selection depends on many factors, like type of alternatives, areas affected by each alternative, project importance, data availability, etc. Therefore, it is probably pertinent to use as many criteria as possible to make sure that everything is covered. However, this is not easy either, because the load of computation works and especially because the monumental task of data collection, which perhaps is not used.

A project or plan may affect many different areas, and criteria must consider them all. A very important feature is to incorporate in the scenario those criteria that do not possess quantitative values, but subjective ones such as '*comfort*', '*easiness to do something*', '*intangible damage*' (such as that produced by the partition of an inhabited area by the construction of a highway), or those that generate externalities. i.e. costs that don't have a market value'.

Impacts produced by projects must also be analyzed in greater detail because, the same as criteria, they can affect many different areas. Especially taken into account in this work are impacts that provoke or influence others like a domino effect.

Criteria can also pertain to diverse areas. For instance, from the environmental point of view, criteria can be further broken down in accordance with several points of view, as for instance:

- *Legal criteria*, that is, the way alternatives comply with legislation. As an example, in a project for a landfill construction, does the project comply with the legal requirement about type of soil, thickness of the lining, or drainage piping to evacuate leachate?,
- *Hazardous criteria*. It refers for instance to industrial discharges into a river of toxic substances that can affect aquatic life, such as fuels, chemicals, etc.,
- *Harmful criteria*. Refers to the discharge of dangerous products into the air such as SOx,
- *Frequency criteria*. For instance, there could be a manufacturing plant discharging hot water into a river. It is not the same if the water is continuously discharged, as if it were discharged at certain intervals and for limited periods,

- *Residuals criteria*. There are many industrial plants in which residues have application for other industries. For example in the case study proposed in Sect. 5.1, one of the alternatives proposes to utilize a fluidized bed boiler. This is a boiler equipped with a bed of calcium oxide which, when combine with SOx present in the fuel, yields gypsum, of commercial value, or may be a power plant that can use its hot water discharge from its condenser for heating applications, etc.
- *Risk criteria*. Normally, there are safeguards in the evaluated alternatives to prevent accidents, such as personnel safety, accidental release of contaminants, geological unknown characteristics of terrain, or for delays, budget exceedence, costs variance, etc. This is a very important kind of criteria and is present in almost every project. Calculation of risk involves a rather complicated calculation; for that reason, and since this pertains to data collection which form part of the strategy for a decision-making process, is that its calculation has been exemplified for a project in Sect. 9.2.2. Because risks may involve doing extra work for an alternative or taking additional measures to preserve safety or for whatever other reasons, these extra funds needed for a certain project can be calculated and placed as scores for each alternative. The independent term (See Sect. 5.1.3) for this risk criterion could be for instance the risk that the DM is willing to accept.

2.6.1 Subjective Criteria

Up to now, it was assumed that scores, i.e. the quantities expressing the contribution of each alternative to each criterion, were real values. In this way, for instance, it can be said that project A, which is a conventional coal burning electric power plant, produces 0.6 kg of CO_2 per each kW-hr generated, while another version equipped with a CO_2 absorption system produces say 0.15 kg of CO_2 per kW-hr generated. These values can be compared and give a clear indication of the degree of contamination that each plant generates.

However, there could be a criterion with subjective values, for instance one to measure '*Degree of satisfaction*' of workers labouring in plants A and B. A worker could say "*It is more satisfying for me to work in plant A, because I am familiar with it*". The problem is how to use this linguistic estimate, which is not a score but an opinion. Model AHP (See Sect. 3.2.4), utilizes a dedicated scale to quantify these preferences. When comparing both projects on the mentioned criterion, the worker could say, pressed for more precision, "*I have a strong preference for plant A*"; and in AHP scale it means a value of 5. This is called '*pair-wise comparison*', and uses a ratio scale, since it is expressing that from that point of view or criterion, A is 5 times preferable than B, and inversely, B is only 1/5 of A.

There are also interval scales where there is no relationship between aspects measured. For instance, a similar question about quality of food could merit 9 in one restaurant and 3 in another; however, it is not possible to say that from the point of view of food quality the cooking in the first restaurant is three times better than in the other. It simply shows that in an arbitrary scale the food in the first restaurant gets almost the maximum and in the second a mediocre mark, nothing else.

By the same token, if three similar cars are compared, it is not right to say that model A is twice as comfortable as D but three times less comfortable than G for it does not make sense, because how do we measure comfort? The same argument applies to certain political projects that have to be considered and analyzed. If out of six alternatives or projects, two are political, how many times is one more important than the other?

Another method does not compare projects vis-à-vis on a certain criterion but simply gives each one a value in a certain scale. In these cases an ordinal range of values is established, for instance '*bad*', '*ordinary*', '*good*', '*very good*', and '*excellent*', and then a metric is assigned to each of these adjectives in a scale say from 1 to 10. The cardinal ordering could be:

Bad = 1, Ordinary = 2, Good = 5, Very good = 8, Excellent = 10

Whatever the method, it is evident that these subjective or ordinal values have to be converted to cardinal ones. These very simple examples show why decision-making based in personal assumptions can lead to different results depending on which person is doing the assessment, which really doesn't make much sense, to say nothing of reliability.

This book includes SIMUS methodology for the selection and evaluation of alternatives. At the very beginning, the method introduces a form to be used in a survey, which contains questions for each project such as '*How does this project affect people*?' *What adverse and favourable impacts produce*? And so on. It is not easy to appraise these queries; however, a logic structure can be adopted to quantify impacts as mentioned.

2.6.2 *Different Criteria Types*

There are different criteria types as follows:

Technical

These are the criteria related with the technical characteristics of each alternative or project. Assume for instance that the problem calls for the selection of four different processes pertaining to different methods to reduce contamination in water discharged from industrial plants into a river. One assessment criterion relates to measuring the BOD_5[10] in the wastewater stream. Analysis can find for instance for alternative B a value of 175, which indicates that alternative B produces an effluent with a contamination of 175 mg/l. Another alternative, say D, shows that for the same criterion the effluent would have a lesser value of about 152 mg/l, and so on. Obviously these values are not meaningful if a *limit* or *threshold* is not established, since the fact that the value 152 is lower than 175, even if it is true that it indicates a lower contamination, *doesn't necessarily mean that this lower value is acceptable.*

[10] BOD_5. Biological Oxygen Demand; indicator used to gauge the quality of effluents.

In this context, our threshold could be for instance 160 mg/liter, consequently, as per this criterion an alternative producing an effluent with a content of more than 160 mg/liter does not match this restriction. With this threshold, it is possible to assert that alternative D is better than B, for the first has lower content than the maximum allowed, while the second alternative surpasses it.

Other criteria relate to *capacities*. For instance, in a project that utilizes raw water and treats it for further use in a water treatment plant, a criterion could specify the maximum capacity of this plant, which naturally, cannot be surpassed.

Availability
From this point of view these criteria refer to the characteristics of some intervening elements such as:

- Availability of the necessary manpower in trades and quantity (professionals, staff, technicians, clerks and hands),
- Availability of equipment, such as loaders, excavators, cranes, large trucks, etc.,
- Storage capacity for warehouses, vessels, lagoons, etc., measured in m^2 or in m^3.

Environmental
Here, criteria usually relate to thresholds *that must not be exceeded.* The following list shows examples of environmental criteria and origin and sources of contamination (in brackets), due to projects:

- Aquifer contamination (percolation from tailing ponds in mining operations, or from city landfills),
- Preserved areas crossing (laying tracks for a railway on saline plains),
- Biological effects (dust from many activities covering vegetation and disturbing the photosynthesis process),
- CO discharges (heating systems, solid waste burning, road vehicles, detonation of explosives),
- CO_2 discharges (power houses, factories, cars),
- Crossing degraded forest (roads, pipelines, transmission lines, aqueducts),
- Crossing native forest (roads, pipelines, transmission lines, aqueducts),
- Crossing natural parks (roads, pipelines, transmission lines, aqueducts),
- Desert crossings (road building),
- Discharges to atmosphere (power houses, factories, cars),
- Dust production (construction works),
- Effect on marshes and crossing marshes (road construction, transmission lines),
- Energy generation from fossil fuels (power houses),
- Fauna migration (decreasing habitat size, noise, lack of food or water),
- Flora and fauna sanctuary (noise, road construction, logging),
- Hazardous wastes (hospital wastes, wastes from chemical plants),
- Hot water discharges (powerhouse condensers),
- Impact on aquifers (extracting water through pumping),
- Land use (housing developments),
- Noise production (airports, gas turbines, wind power),

- NOx discharges (power houses, factories, cars),
- Number of km of visible works (oil pipeline on surface, not buried),
- Number of trees to be logged (forest industry, roads, transmissions lines, railways),
- Number of river crossings (road construction, pipelines, railways),
- Particulate discharges (power houses, factories, cars),
- Phosphorous (sewage treatment plants),
- Swamp crossings (roads, pipelines, transmission lines, aqueducts),
- Toxic discharges into water sources (industrial plants such as plating shops),
- Visual contamination (advertisement boards along a road or in mountains),
- Water uses (industrial process, irrigation, potable water),
- Wetlands crossings (roads, pipelines, transmission lines, aqueducts),
- Etc.

In most of these projects thresholds can be established for criteria.

Risk
Regarding risk, some criteria deal with thresholds expressed in percentages. As an example it is possible to say that soil contamination risk is possibly too low at 2%, or that the risk of social unrest is relatively high on the order of 15%. The multicriteria analysis model allows examining these cases, and if this were the only criterion, the model possibly would select the lesser risk alternative. However, usually this is not the case and the model has to find a compromise between the different criteria types related with:

- Energy risk,
- Geological risk,
- Political risk,
- Sabotage risk,
- Safety risk,
- Seismic risk,
- Social unrest,
- Soil contamination risk,
- Etc.

Social criteria
It is usual to utilize percentages for social criteria, expressing *perception of people about different subjects*. For instance, a project may receive 45% public approval while another receives 70%. Social criteria examples are:

- Citizenship evaluation and opinion in favour or against a project,
- Crime during construction,
- Floor space in m^2 per person in housing projects,
- Heritage conservation,
- Minimum quantity of water per person and per day,
- Number of inhabited areas which are at a certain distance from the project site,
- Number of people affected by a project and how they can be affected, for instance relocation because a public development,

- Number of people to be relocated,
- Prevailing winds as in the case of the construction of a domestic waste incinerator,
- Public acceptance,
- Public health and safety.

Economics
These are criteria expressing *economic aspects*, as for instance the number of hectares of cultivated fields, or hectares of crops that are affected by an alternative, or the vehicle traffic expressed in vehicles/h. Examples are:

- Commercial forests,
- Cultivable land,
- Cultivated fields,
- Direct economic benefits,
- Economic efficiency,
- Efficiency and resources use,
- Indirect economic benefits,
- Market competition,
- Merchandise flow,
- Merchandise volume,
- Production unit cost,
- Urban movement,
- Etc.

Construction
These are technical criteria like the slope of a terrain or the length of a road. Examples are:

- Difficult access to site,
- Distance (for instance, transmission lines),
- Geological faults,
- Geological stability (or lack of it),
- Lack or scarcity of water,
- Logistics,
- Open pit works, as in mining,
- Technical difficulties,
- Type of soil and suitability for projects, for instance places where landfills can be built, because it is necessary to have a clayey soil, Etc.

Spatial
These criteria relate to *spatial effects* for an alternative or project. If the project extends spatially, the effect in distance can be expressed here. Examples are:

- Diffusion of a contaminant due to winds. Typical examples are the exhaust gases from a domestic waste incinerator,
- Minimum distance to an inhabited area,
- Rights of way,

- Trans-boundary effects, as is the case when the impacts are felt across borders between countries. Typical example is acid rain produced in U.S.A. with effects in Canada, Etc.

Temporary
Criteria related to persistence of an effect. Examples are:

- Removing trees from an avenue to make room for underground construction; once completed, trees will be replanted,
- Working camps for large construction projects (for instance paving in the office and camp area which will be removed after the project is completed),
- *Length of time* considered for the effect.

Cultural, for instance:

- Project impact that implies a *change of life* condition,
- Project impacts when crossing areas that for whatever reasons are considered of *cultural*, *religious* or *heritage* importance by a group of persons,
- Projects developed nearby archaeological zones, which somehow can *damage* them either with emissions, vibrations or with any other effect.

Legal. That is, criteria that are related to laws, regulations, human rights, etc. An example is:

- A large and very well-known company developed a hydro project that implied rerouting water from a lake located high in a mountain, and which native people used for fishing. These people estimated that water extraction from the lake would *damage their activity* because it would provoke a decrease in the lake level, with a direct effect in the river flow that evacuates the lake, and from where the natives fish trout for food. Consequently, they initiated a legal action against the company building the hydro scheme and that apparently did not take adequate notice of this claim. This actual case happened in Canada, and finished when the Supreme Court ruled discontinuing the project in 1991, because it understood that the project violated native rights in the area. This project involved about 1,500 million dollars and when stopped nearly 40% had been executed, and it continues closed until to this day.

Financial
Generally refers to project financing and usually address aspects like these:

- Accepted p*ercentages* of change or variation of prices, fixed costs, acquisition of equipment, working capital, etc.,
- Criteria used for project profitability, such as *Net Present Value* and/or the *Internal Rate of Return* of each one,
- Influence of each project in the *balance of payments* of the country,
- *Maximum funds availab*le to develop these projects,
- *Minimum and maximum amounts* which can be devoted to each project,
- *Origin* of funds,
- *Periods* when funds will be available for these projects and amount per period.

Infrastructure
Aspects to consider

- *Construction technical difficulties* due to ground or soil conditions, or perhaps lack of water, or transportation, as it could be for instance in a mining project in a mountainous area where a route must be available to allow traffic of heavy trucks,
- *Logistics, i.e.* ease in getting inputs, through harbours, airports, railway nodes, etc. For instance in a large mining project in Peru it was necessary to make some changes in the closest maritime harbour to allow disembarking heavy equipment. In this same project, it was necessary to reinforce bridges along the 320 km road to allow the transportation of this heavy equipment,
- Traffic *flow*,
- Traffic *volume*.

2.6.3 Fields Covered by Criteria

Since criteria are the elements used to evaluate alternatives or projects, it is important to check all fields or spheres of action affected by their potential impacts; as an example consider the selection of location alternatives for a project such as the construction of an aluminium smelter. The plant, which is possible to build in different places in the country or in the region, or even in other countries, will have naturally, criteria related with the economic aspect, but there are also other fields or areas, which are as important as the economics. Most probably, the factory will be on a maritime shore to allow ocean vessels to supply the plant with its raw material, bauxite,[11] and in large quantities, because production of 0.5 kg of metallic aluminium requires 2 kg of bauxite. This fact most probably will demand the construction of an industrial wharf and the laying of railway tracks or perhaps conveyors and cranes. This is heavy work, and most probably it will alter the marine ecosystem because of the water disturbance produced by ships and the potential shifting of sand created by piloting the wharf.

The aluminium smelter by itself will produce atmospheric contamination, most probably affecting the natural vegetation of the area by noxious emissions from the plant. It is possible to neutralize this effect to a certain extent, but adopting measures in the initial phases of the project, by studying different alternatives and safeguards, and then establishing thresholds of tolerance. On the other hand, aluminium smelters are large generators of tetrafluoromethane and hexafluoroethane, recognised culprits of part of global warming, and therefore, adequate measures have to be taken to decrease the amount of these emissions.

Almost with certainty the plant will hire personnel from the area or, which is most probable, will attract them from other areas. Nothing wrong with this, but it means housing construction, the supply of basic infrastructure (water, sewage, electricity, etc.), as well as amenities, recreation centres, etc. It is unnecessary to

[11] Mineral composed of about 45–60% of aluminium oxide (O_2Al).

mention that the project must calibrate the negative effects to the population produced by the discharge of potential noxious emissions – which can also be accidental – and the release of the above-mentioned gases to the atmosphere. Therefore, it is very important to take into account in the planning stage which criteria to use to evaluate alternatives considering these serious effects.

Odour is a component that usually is sub-evaluated or even not taken into account, or not well estimated. For instance, a large meat packing plant, a wastewater treatment plant, a pulp mill, etc, can produce permanent odours, which without a doubt will create future problems for the nearby population.

From the above it is clear that a very important aspect when identifying the criteria to be used in the evaluation, is to consider all the areas or fields affected by the different projects.

2.6.4 Criteria Weights

In many multicriteria projects, the DM and staff assign a weight to each criterion; there are several ways to obtain these weights. Most of them use the opinion of experts who compare the relative importance of each criterion against the others. All of these methods entail subjectivity in assigning weights to criteria, and because of that, there is no guarantee that these weights will be replicated when another person or team estimates them, and of course we are talking of the same set of projects and under the same conditions and assumptions. There are however more realistic mechanisms to weight criteria; professor Milan Zeleny, a U.S. mathematician, designed an elegant and effective method to weight criteria in which there is no subjectivity because weights come from the alternatives scores.

To illustrate it let us consider several alternatives and say five criteria. Alternatives are placed in columns while criteria in rows. There is always a criterion in each row and thus, a criterion evaluates all alternatives.

For each criterion, there is a score from each alternative, placed at the intersection of the alternative and the criterion. This score expresses how well each alternative contributes to the objective expressed by that criterion. Generally these values can be normalized and then take values between '0' and '1'.

Sometimes, and it is very usual, there are large discrepancies between these scores considering a criterion in particular, that is, there could be large variations between these numbers on a criterion when all alternatives are considered. For instance, suppose that in a mining project there are three alternatives to transport ore, and from the point of view of criterion 'Cost', these are the normalized values:

(C_1) 'Cost' criterion 0.58 0.94 0.32

And there is also a criterion such as 'Net Present Value' with these normalized quantities:

(C_2) 'Net Present Value' Criterion 0.85 0.88 0.83

Note that 'Cost' criterion displays a larger discrepancy in values than 'Net Present Value' criterion.

From the criteria significance point of view it is better to have a large discrepancy because it means that the criterion can *discriminate* or differentiate amongst various alternatives. To measure this degree of discrimination Zeleny employs a concept developed by the American mathematician Claude E. Shannon (1948), in his well-known work '*Mathematical Theory of Communication*'. In this famous book Shannon, considered the father of modern Information Theory, established a measure of the information content in a message, which he called '*Entropy*', which also is a well-known function in Thermodynamics, a field in which it registers, in a certain measure, the level of disorder.

Shannon's formula is:

$$\mathrm{H(entropy)} = -\mathrm{K}\sum_{\mathrm{i}=1}^{n} p_i \log(p_\mathrm{i})$$

K = Constant, which depends on the unit of measured selected,
p_i = Probability of an event occurrence.

Zeleny (2000) applied this concept but used score values in lieu of probabilities, and thus, utilizing this formula for each criterion, determined which of them provides the *maximum quantity of information*, which allows for establishing a measure of its importance, or weight. Obviously, if on a given criterion all alternatives or projects have similar scores, then that criterion will provide very little information since all the alternatives contribute with the same intensity; in this case, most probably the criterion is useless for evaluation. Considering that the lower the entropy the better, if this formula is applied to the above example the result is:

$$\text{Entropy } (\mathrm{S_1}) \text{ for 'Cost' criterion} = -0.246$$

$$\text{Entropy } (\mathrm{S_2}) \text{ for 'Net Present Value' Criterion} = -0.135$$

As expected, the 'Cost' criterion is more significant to evaluate alternatives than the 'Net Present Value' criterion, regarding the information content, since:

$$\mathrm{S_1} = -0.246 < \mathrm{S_2} = -0.135.$$

This writer has also used this entropy concept but with the purpose of *selecting alternatives offering the maximum quantity of information* when the method is applied to environmental indicators selection. Barba – Romero (1997) expresses, regarding this methodology that:

> A completely different way is the Entropy method (Zeleny 1982). The essential idea is that the relative importance of criterion j (to be measured by weight w_j) in a given situation of decision (referred to the decision matrix) is directly related with the average intrinsic information generated by the set of alternatives regarding each criterion, and also by the subjective assignment of the importance given by the DM according to his/her preferences

Linear Programming also computes criteria weights but without subjectivity, and this information appears automatically when a problem is solved. In this case the values assigned to each criterion represent its imputed values and also correspond to what is known in Economics as '*Shadow prices*' or '*Marginal values*'.

2.6.5 Examples of Threshold Standards, Types and Units of Measure

There are international standards in different fields and used as guidelines. These values are very useful when it is necessary to quantify thresholds for different criteria. The following short list enumerates some international standards and their units of measure.

- *Air, soil and water contamination* (SOx, NOx, CO, CO_2, BOD_5, P, N), in their appropriate units of measure, ppm (parts per million) mg/l, ppb (parts per billion), etc.,
- *Capacity of public services* like hospitals and police (number of people/bed, average waiting time for surgery, number of people/ambulance, crime rate, break-in rate, etc.),
- *City management* (number of City Hall employees/city population, municipal personnel salaries and wages/city budget, city budget managed by the community/total city budget, percentage of metered residual water use, city passenger travel by mode, ratio people/cars, etc.),
- *Environment* (green space in m^2/person in a city, effluent treated in water treatment plants/total effluent, percentage of paper, glass, metal and plastic recycled per year, average annual ozone levels, etc.),
- *Financing* (interest rates in percentage, payback periods in years, Internal Rate of Return in percentage, etc.),
- *Landfill protection against leaks* (type and thickness of plastic lining),
- *Maximum capacity in public service installations/inhabitant*, such as raw water treatment plants and waste water treatment plants,
- *Risk* values, in percentage (construction industry, manufacturing, road construction, etc.),
- *Urban indicators* (percentage of dwellings connected to sewage, potable water per person/day, percentage of dirt streets, percentage of university students related with city population),
- Etc.

There is abundant literature of tables with values for the concentration of gases in clean air, which can be used as standards, such as in (Seinfeld and Pandis 1998).

There is a good example of using these thresholds in Monitoring Report Technological and Environmental Management Network Ltd. (2002).This publication shows how the *actual values* from a dredging operation compared with *standard thresholds* established for each criterion.

Table 2.2 indicates the main origin of some contaminants, which can also be used as a first guide when analyzing a project.

Table 2.2 Main origins of some contaminants

Pollutant	Chemical formula	Most common origin
Nitrogen gases	NOx	Cars, electric energy generation plants
Sulphur gases	SOx	Electric energy generation plants burning coal with a high content of sulphur. Oil refineries
Carbon monoxide	CO	Idling cars
	Hydrocarbons	Gasoline and gas oil operated vehicles Farms producing methane gas from Animals digestive tract
Various gases	Smog[a]	Mainly from vehicles exhaust combined with solar light
	Particulate	Electric energy generation from plants burning coal Industries
	Dust	Construction works. Atmospheric action
Lead	Pb	Car burning leaded gasoline

[a] Smog. This word identifies a blend of contaminants with ground level ozone as main component. Ozone is formed through a photochemical process when nitrogen and volatile compounds interact with solar light

Sometimes criteria relate to each other. In this circumstance, it could be that some of them are redundant. The use of Factor Analysis (FA) can help in eliminating redundancy.

2.6.6 Magnitude of Thresholds

As explained, a threshold is a metric, used *to delimit* or *to establish limits to a criterion*. In this way thresholds are sort of *benchmarks* to show that a criterion is limited in its scope. Many criteria use thresholds that are in reality indicators, for example, if a criterion relates to water consumption, it is then possible to assign a threshold value of 255 l/person-day, which is the international standard. However, depending on the zone, there could be another threshold value.

By the same token, a threshold can represent some measure of *sustainability*; for example the rate at which an aquifer is naturally recharged. This value shows then the *carrying capacity of the environment*, i.e. reports about the continuous capacity of the environment to sustain life. Table 2.3 shows some threshold values commonly used in projects.

2.6.7 Examples of Thresholds and Characteristics

Some examples of thresholds regarding:

- Capacities
  In a hydroelectric project it is necessary to work with the potential capacity in each site to store water in hm^3, with lower and upper limits in meters, and with

Table 2.3 Examples of threshold units

Area	Units of measure
Environment	
Domestic solid wastes	kg./person-day
Maximum CO content in streets in an 8 h period	$\mu g/m^3$
Paper recycling	%
Particulate	mg/m^3
Infrastructure	
Flooded streets in heavy rainfall	%
Dwellings connected to water network	%
Vehicles flow	vehicles/h
Transportation	
Investment in road infrastructure	$/capita
Social	
Total number of dwelling units	houses/1,000 persons
Average floor space per person	m^2/person
Government	
Ratio between wages of government personnel and local expenses	%

values for permeability, average rains, floods, etc. That is, it is not enough to estimate that site A has potential for electricity generation of say 400 MW and B a capacity of 500 MW, and C 250 MW, since it is necessary to establish thresholds according to needs. If these needs are for instance as a minimum 1,500 MW none of these alternatives are feasible.

- Demand and economic level
  Consider as an example a large housing development project, with two types of houses. It is normally not enough to decide how many of each type to build, based only on the assumption that because houses of type A have 40 m^2 more floor space than houses of type B, they will be more comfortable considering habitability, and will sell faster. It is necessary too to establish thresholds reflecting people's actual needs, and the existent selling potential for each type of house. Another fact is the necessity of estimating the average economic position of potential purchasers. Why? Because, if not, the entrepreneur could build 278 houses of type A and fewer houses of type B, and then discover that there is not enough market for the larger houses (or because they are too expensive for the average buyer), and instead there is a scarcity of B dwellings, or vice versa.
- Limits for allowable contamination
  If diverse energetic projects are considered and analyzed from the environmental point of view, say regarding NOx production, it is futile to say that project C is better because it generates less contamination than project F, since this appreciation depends on the allowable limits for that type of contamination.
- Priorities
  Assume the following example: In urban road infrastructure, on what basis does the DM select laying down pavement for district J or for district H, when

budget restrictions allow for funding only one of them? Naturally, it is possible to establish this comparison based in linear metres of dirt streets. However, since dust produced by traffic affects people's lungs, there is need for additional kinds of data, such as:

(a) What amount of dust in air measured in grams/m^3 is detrimental to human health?
(b) How many people live along that street? and
(c) How much traffic is there in each alternative?

When the model is loaded with this kind of information for both districts, then it is possible to establish a priority.

- Requirements
  There are projects that require a minimum level or volume, as for instance projects to impart instruction on several trades. There must be a minimum number of students, for economical reasons, and a maximum number to maintain the ratio of professor/students at an acceptable level.
- Quantities
  Sometimes the DM specifies that there must be a certain number of alternatives complying with a particular criterion
  This is the case when environmental indicators are analyzed and where a certain minimum number of indicators per criterion is demanded. See example in Sect. 8.3.2.
- Cash flow
  In large projects, alternatives may involve building different ancillary structures in different times. In those cases, it is wise to add such criteria as annual periods and specify funds available for each period. For instance, assume that three projects start at different times and have different durations. Project A will take 5 years; project B, 3 years, and project C, 1.5 years. There is an annual budget for each fiscal year and the schedule specifies the percentage of each project to complete each year. See Table 2.4.

Table 2.4 Schedule of completion and funds availability

Projects	A	B	C	
Fiscal year	Percentage of completion per fiscal year	Percentage of completion per fiscal year	Percentage of completion per fiscal year	Available budget (Euros)
2011	0.12	0.05		2,050,631
2012	0.35		0.60	1,896,241
2013	0.43	0.65	0.40	760,000
2014	0.10	0.30		826,741

It is evident that in 2011 the earned value (that is the work actually done in monetary terms) for projects A and B must not exceed 2,050,631 Euros, and the same analysis is valid for other years. If this is not specified it could very well be that funds assigned for each year are not enough in 1 year and in excess in other years.

2.7 Impact Evaluation

2.7.1 How Are Impacts Evaluated?

This is a difficult task and sometimes impossible, especially considering that an impact is able to trigger another effect, which in turn unleashes a third one and so on; but if we know for a project the effects that provoke an impact, it is often possible to determine which are most important. That is we need to know what causes the impact, and in so doing perhaps it is possible to reduce it. For instance, consider an open pit mining exploitation which broadly speaking consists of: (a) Mining the mineral, (b) transportation of ore to the crushing centre, (c) chemical process such as floatation to extract the mineral, and finally (d) disposal of floatation residues which are called '*tailings*'.

Figure 2.1 displays three partial aspects of this process, which involves ore transportation, crushing and milling or grinding (to reduce the mineral to a fine powder able to be attacked by the chemical solution in the float process).

These three activities generate contamination such as road dust from transportation and ore dust from the other two actions. This dust, either from one origin or the other, affects persons, animals and vegetation (in the latter reducing the absorption of sun energy which is essential for the photosynthesis process which delivers oxygen to the atmosphere).[12]

Here, if a value of 7 (in a 1–10 damage scale, the larger the worst), is the qualification of impact from 'Transportation' in dust generation and the influence of this dust on persons (Receptors) is estimated as 4, then:

For PERSONS as receptor:

Impact of Transportation = 7 + 4 = 11
Impact of Crushing = 2 + 4 = 6
Impact of Grinding = 3 + 4 = 7

For WATER as a receptor:

Impact of Transportation = 7 + 8 = 15
Impact of Crushing = 2 + 8 = 10
Impact of Grinding = 3 + 8 = 11

For AIR as a receptor:

Impact of Transportation = 7 + 8 = 15
Impact of Crushing = 2 + 8 = 10
Impact of Grinding = 3 + 8 = 11

[12] It is perhaps difficult to assess in its real value the impact caused by dust in plants. To this effect it is interesting to replicate what was reported by the Las Provincias newspaper in its August 30, 2009 edition when it commented about the damage that the workings for the water transfer scheme between Júcar and Vinalopó rivers, Spain, is producing in crops in the area, which states "*Orange trees are choking and are whitish because of the dust and because the damage from the dust impedes the photosynthesis*".

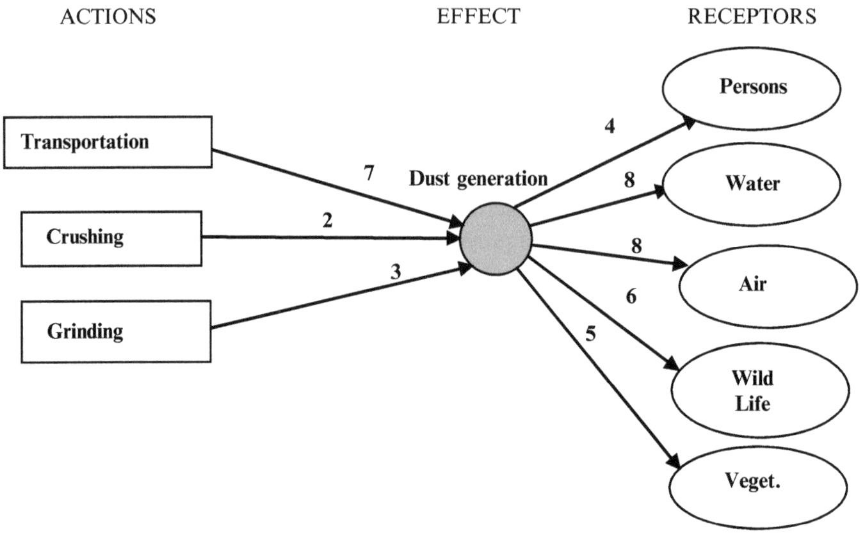

Fig. 2.1 Relative importance of effects produced by dust

For WILDLIFE as receptor:

Impact of Transportation = 7 + 6 = 13
Impact of Crushing = 2 + 6 = 8
Impact of Grinding = 3 + 6 = 9

For VEGETATION as receptor:

Impact of Transportation = 7 + 5 = 12
Impact of Crushing = 2 + 5 = 7
Impact of Grinding = 3 + 5 = 8

As a summary, adding up impacts for each action:

Transportation = 11 + 15 + 15 + 13 + 12 = 66
Crushing = 6 + 10 + 10 + 8 + 7 = 41
Grinding = 7 + 11 + 11 + 9 + 8 = 46

2.7.2 Criteria Definition

Obviously, transportation originates the largest effect, followed by grinding and crushing. With this procedure we have detected the main cause and evaluated the relative importance of actions that generate dust. Once this is known, three sequential mechanisms can be implemented to address them, as follows:

(a) Take measures to decrease dust generation.
These measures can include decreasing truck speeds, water spraying vegetation at the edges of the road, improving better fitting in hatches lids built in crushers and grinders, supplying crusher and grinder operators with dust masks, etc.

(b) Define criteria related with dust generation.
Some criterion could be related for instance with transportation. In this sense a criterion could establish that truck speeds must not exceed a certain road speed limit in specific parts of the service road. Other criteria can be established for crushers and grinders related with permissible dust content in atmospheric air around these installations.
(c) Establish thresholds for these criteria.
Thresholds could be then set up for instance at 45 Km/h in the road and reduced to 30 Km/h in certain portions of it, like any other road signals. For dust concentration in air values such as 0.5 mg/m^3 can be adopted.

Dust generation may perhaps be appraised as a minor or a little significant effect, and disregarded because this judgment, however, it can influence project production, because of the necessity to eventually reduce speed in heavy trucks, and with that the quantity of ore transported to crushing and grinding, which in turn decreases metal concentrate production in the floating process, and as a bottom line, less plant production.

In another project, such as improving life conditions for people living in 'shanty towns' assume that there are two alternatives, (a) relocating people into new houses built on the outskirts of the city, and (b) keeping the people in the same place but trying to integrate it into the city fabric through the provision of basic infrastructure. It is obvious that there must be a criterion such as '*Disruption of way of life*' with scores from both options, which must be minimized, as well as other such as '*Integration to city fabric*', that must be maximized. The type, number and scope of criteria depend on each project and in the depth the DM wants to reach in this analysis. On the other hand, most computer models have a maximum number of criteria, not only from the software point of view, but mainly considering the volume of information that has to be collected.

2.8 Conclusion for This Chapter

This chapter aims at giving practical information about data collection and its processing, a fundamental part of the decision-making process. From this point of view some principles in relation to procedures have been commented on, as well as suggestions about the necessity of making sure that people at diverse levels will be engaged. That is, those members of the decision team and people who will be the recipient of its benefits or suffer some of its inconveniences. Later on, projects and alternatives were examined in depth and their relationships and different types of projects commented on. The fundamental criteria concept has been thoroughly examined in its various aspects, including weights, as well as thresholds, and facts and intelligence have been given about international standards, as well as tables with data of main origin of contaminants related with diverse types of projects. Orientation was also given on units of measure in diverse fields or areas. Impacts in turn, are also studied and a methodology is suggested for their evaluation.

In summary, this chapter gives information about know-how and how to collect and analyze relevant data which will be used in Chap. 4.

References

AALBORGF+10. *Inspiring future*. Retrieved September 23, 2009, from http://www.aalborgplus10.dk/media/finaldraftaalborgcommitnts.pdf

Barba-Romero, S. (1997). Conceptos y soportes informáticos de la decisión multicriterio discreta. Fuente: Martínez, Eduardo y Escudey, Mauricio (Eds.), *Evaluación y decisión multicriterio: reflexiones y experiencias*. Santiago: USACH/UNESCO©.

Seinfeld, J. H., & Pandis, S. N. (1998). *Atmospheric chemistry and physics hierarchy of air pollution*. New York: Wiley.

Shannon, C. (1948). Mathematical theory of communication. *The Bell System Technical Journal, 27*, 379–423–623–656.

Stolp, A., Groen, W., van Vliet, J., & Vanclay, F. (2002). Citizen values assessment: Incorporating citizen's value judgments in environmental impact assessment. *Impact Assessment and Project Appraisal, 20*, 11.

Zeleny, M. (1973). *Multiple criteria decision*. New York: McGraw Hill, 1982.

Zeleny, M. (2000). In Y. Shi & M. Zeleny (Eds.), *New frontiers of decision making for the information technology era*. Singapore: World Scientific Publishers.

Chapter 3
State of the Art in Decision-Making

Abstract The purpose of this chapter is to have a look at the most current models and methodologies used for helping the DM. It is not a detailed analysis because there is abundant bibliography on each one of them; it gives instead enough information to learn about their capabilities, limits and potential, and thus enabling the DM to choose the model that he/she believes is more adequate. This chapter examines the five most popular models, which are MAUT, ELECTRE, PROMETHEE, TOPSIS and AHP, and comments on an expansion of the latter known as ANP. However, it does not consider Linear Programming because its explanation and exemplification are in Chap. 4. It is a common belief that there is no one method superior to another, albeit there is perhaps one that is more popular, but most of the time any of them can be used to solve a problem; however, there is a comparison made on their characteristics and that is illustrated in Chap. 7.

Keywords State of the art • Decision-making • Models • Model's examples • Decision matrix

3.1 Most Used Methods for Helping Decision-Making

3.1.1 General Information

> *It has become more and more difficult to see the world around us in a one-dimensional way and to use only a single criterion when judging what we see.*
>
> (Milan Zeleny 1982).

By the end of the 1960s, and due to work in France by Roy (1968), decision-making developed in establishing models such as the ELECTRE family. They are based on comparing paired projects or alternatives; the dominant is chosen in accordance with certain principles or norms and the dominated discarded, until a satisfactory

N. Munier, *A Strategy for Using Multicriteria Analysis in Decision-Making: A Guide for Simple and Complex Environmental Projects*, DOI 10.1007/978-94-007-1512-7_3,

solution is found; for this reason these methods were given the name *'outranking methods'*. Roy developed his theory on the grounds of considering concordance and discordance levels, which allow determining under what circumstances one alternative dominates other. In addition, many American researchers have worked on this issue (See Lootsma 1999).

Years later a method was developed called PROMETHEE (Brans and Vincke 1985), an example of an outranking model in which transfer functions replaced the above-mentioned levels, using preference thresholds, and associated with the GAIA graphic system. There are other methods like REGIME (Hinloopen et al. 1983), (Munda 2004), MAUT (Wallenius et al. 2008), and Saaty (1980) with his *'Analytical Hierarchical Process'* (AHP). The latter also contrasts paired activities, albeit in a different way because it uses DM's preferences, but without a doubt they all greatly contributed in clarifying the issue and in offering help for the decision-making process.

One of the first methods was MAUT (Multi Attribute Utility Theory), based on utility theory[1]; it has had considerable success especially in the United States. It is an additive method consisting in multiplying the score for each alternative and for a criterion, by the weight assigned to that criterion. Further, it proceeds with the summation of values found; the selected alternative is the one that gets the highest value from this summation. According to Vincke (1992), the purpose in developing MAUT was to take into account uncertainty caused by lack of precise information or data; consequently, the model uses probabilities, in which case the probability of occurrence substitutes for the weight.

Other methods, such as TOPSIS (Hwang and Yoon 1981), find solutions considering the distance that separates each alternative from the ideal solution. The shorter this distance the better the solution, and this is the comparison that allows us to determine the dominance of one alternative over other.

However, it is necessary to mention that all these techniques deal with only partial aspects of the problem, that is they suggest the most suitable alternative amongst several and can also show a ranking of alternatives, but they do not ensure that the solution found will be the best. By contrast, the Linear Programming technique, with one objective, can guarantee that the solution found is the best, or optimum, and which cannot consequently be improved. Very often however, the DM is not interested in finding the ideal solution but a satisfactory one that is a solution acceptable by stakeholders.

3.2 Review of the Characteristics of Outranking and Additive Methods

We do not mean in this review to illustrate the workings of each method, since this is not the objective of the work, but only to comment on their main characteristics and to help the DM understand where their power and convenience lie, as well as to avoid their weaknesses. This book points out the importance and versatility of Linear

[1]Utility. The meaning of this term in the setting of multicriteria decision-making relates to the measure of satisfaction experienced by a person who receives a good or a service.

Programming (LP), and in so doing, intends to show its advantage over other methods. Naturally, LP also has weak aspects and these are also analyzed and discussed.

The following sections illustrate the way each model works. To better explain the procedure there are examples proposed for each one, and some solved by hand and then by the respective software. These 'double' solutions aim at showing for each model its working features and procedures, to allow the DM or analyst to understand the process.

3.2.1 *MAUT (Multi Attribute Utility Theory)*

The formula to compute the utility from an alternative is:

$$V(x) = \sum_{i=1} w_i \cdot v_i(x)$$

Where:

V(x) = Utility of alternative x,
$v_i(x)$ = Value of alternative x according to criterion (or attribute) 'i',
w_i = Weight or importance of criterion or attribute 'i'.

3.2.1.1 Illustrative Example: Choosing an Apartment (Solved by Hand)

A simple example is proposed: A person wishes to purchase an apartment and establishes a series of preferences or dimensions (criteria) which are Habitability, Comfort and Accessibility (transportation to and from). At the same time each criterion has its own qualities or characteristics, which we call *'attributes'*. A *'tree'* involving dimensions and attributes is shown in Fig. 3.1. The person indicates his preferences as A = 0.55, B = 0.35, C = 0.10.

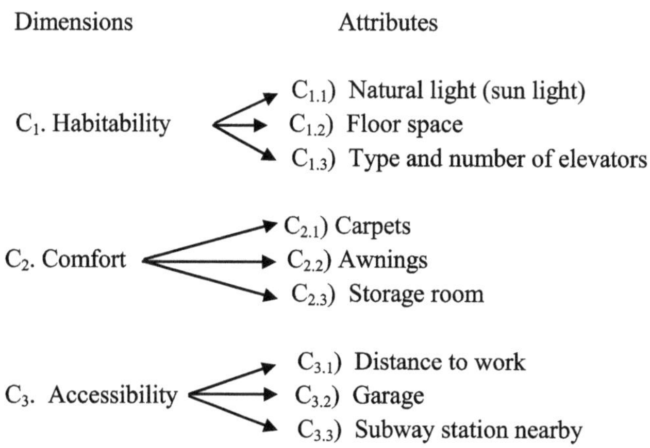

Fig. 3.1 Tree with dimensions and attributes

Table 3.1 shows data corresponding to three apartments. Values are obtained in a 1–10 scale, the larger the better, and the same for weights in a 0–1 scale. For instance, for the buyer, Apartment A has a value of 9 regarding floor space. Regarding weights, the purchaser favours natural light versus floor space and number of elevators, and this preference is the same for both alternatives.

Table 3.1 Values for each apartment regarding habitability

	Apartments			
	A		B	
Attributes	Value	Weight	Value	Weight
$C_{1.1}$) Natural light	$v_a=5$	$w_a=0.5$	$v_a=8$	$w_a=0.5$
$C_{1.2}$) Floor space	$v_b=9$	$w_b=0.35$	$v_b=3$	$w_b=0.35$
$C_{1.3}$) Type and number of elevators	$v_c=4$	$w_c=0.15$	$v_c=6$	$w_c=0.15$
	$5\times0.5+9\times0.35+4\times0.15=\mathbf{6.25}$		$8\times0.5+3\times0.35+6\times0.15=\mathbf{5.95}$	

Weight for criterion C_1, will be: $6.25\times0.55=3.44$ for apartment A, and y $5.95\times0.55=3.27$ for apartment B. A similar procedure gives the values of A and B for each one of the other dimensions C_2 and C_3, which allows for the construction of the Decision Matrix, Table 3.2.

Table 3.2 Decision matrix

	Apartments	
Criteria	A	B
Habitability	3.44	3.27
Comfort	3.84	4.91
Accessibility	2.69	2.43
Total	9.97	10.61

B is the selected apartment because it has the largest valuation. The simplicity of the method is evident and, although it is recognized that many subjective values are applied, it is one of the most accepted. As seen, this method, as others, uses weights; it is worth considering that the weight assigned to a particular criterion may have no relationship with the 'actual' weight or importance pertaining to said criterion due to the dispersion of scores of the different alternatives (See Sect. 2.6.4). As an example, suppose that we are analyzing several projects, and have reached a consensus on the fact that the Internal Rate of Return (IRR) is the most important criterion, and consequently we assign it a high weight, for instance 25%.

However, a more detailed analysis of the IRR expected for each alternative shows that in reality the difference between the lowest and the highest value is small. This leads to the conclusion that the importance of that criterion cannot be as high as thought due to its small discriminatory power; that is, its reduced capacity to differentiate among alternatives, and consequently a small difference in the weight of this criterion in a sensitivity analysis, is not representative and cannot influence the decision. This is the foundation of Zeleny's method for assigning weight in consonance with the dispersion of values when all alternatives or options on a criterion are considered, that is applying the entropy concept as described in Sect. 2.6.4.

3.2.2 ELECTRE (Élimination Et Choix Traduisant la Réalité)

This method partitions the set of solutions into two areas. One of them, called the *'nucleus'*, clusters feasible and most favourable alternatives, and a second one involves the least favoured. The method is obviously focused on the first area i.e. those in the nucleus or *'kernel'*,[2] since any alternative outside that kernel is outranked by an alternative pertaining to the kernel.

ELECTRE belongs to the outranking method family because it establishes a binary relationship between alternatives regarding all criteria (See Rogers et al. 1999). It says that 'a' *outranks* 'b' and is expressed as 'aSb' if the number of favourable criteria favours 'a' and if there is no strong opposition toward 'a'. A *'concordance matrix'* is built, comparing paired alternatives, where outranking exists if there is a strong supremacy in criteria, and there is another *'discordance matrix'*, which opposes the former in the sense that it opposes the supremacy of one alternative over another. Naturally, it comes up immediately as a question about which value we can consider greater enough to justify a supremacy of one alternative over another, which leads to the threshold concept.

To build the outranking matrix, the concordance matrix, the discordance matrix, and the minimum concordance threshold are used, the latter set up by the DM, to accept this concordance and discordance. For each criterion and by each alternatives pair, a concordance index matrix is built, considering the thresholds and the outranking of one alternative over other. There are several versions of ELECTRE, whose general characteristics are:

ELECTRE I: Selects a satisfactory set of alternatives, and works with concordance indexes. They measure the intensity of the arguments favouring the assertion that action (a) outranks action (b). There is also a discordance index, that is, the quantity or intensity of opposed arguments within the criteria under analysis, which challenges the assertion that (a) outranks (b) (Flament 1999).

ELECTRE II: Selects an ordering of alternatives and adds thresholds to the latter matrixes.

ELECTRE III: Similar to ELECTRE II but also adds evaluated outranking relationships and utilizes pseudo criteria, that is attributes which use preference and indifference thresholds.

ELECTRE IV: Similar to those already commented on, however, an important consideration is that it does not require weights for criteria, which is a step further against subjectivity. The interpretation of this lack of weight, is, according to Flament (cited) that *"This doesn't imply that all criteria have the same importance, but that none of them has an inferior category in its relationships with others"*.

This writer does not share this opinion but considers that not all criteria have the same category or importance. For instance, it is obvious that if a project includes

[2] Kernel. Subset of the elements of a set in which a function is transformed to an identity element in another set.

the Internal Rate of Return as a criterion, and as another criterion the payback period, the former is more important than the latter, for it defines the project's profitability, while the second criterion may be negotiated.

As in all models, there are doubts with ELECTRE. Some researchers have reservations and apparently, with reasons, e.g. that fixing thresholds may lead to finding that the corresponding nucleus is empty. This entails modifying these thresholds until a nucleus is generated, which constitutes a manifest arbitrariness or biasness of the system.

3.2.2.1 Illustrative Example. Selection for a Tourism Undertaking (Solved by Hand)

An entrepreneur wishes to take advantage of the possibilities that a certain region offers for tourism the year round (snow in winter and trekking in summer, sailing, fishing, hunting, etc.), as the region is endowed with many natural features that can be exploited for tourism, as mountains, lakes and beautiful scenery. The firm is considering two types of projects (a) A project just to lodge tourists, and (b) projects that create some additional attractions. The different projects considered are:

Project A: Construction of hotel and casino. This project belongs to the first type since it adds nothing; it will only try to attract people by providing lodging and entertainment opportunities, for people to rest, gamble, socialize and enjoy the scenery, and taking advantage of already existent facilities such as walking trails, fishing ponds, horse riding trails, gliding perches, etc.

Project B: Involves the construction of a cable car from the village to the top of a nearby mountain, which offers spectacular views, and building a restaurant at the top. This project then adds a new attraction to the area.

Project C: Construction of a very large sporting and entertainment facility including swimming pools, climbing walls, saunas, bowling, tennis and golf courses, etc. as well as dancing halls, and a mall.

Project D: Construction of a movie complex, complemented with restaurants and discos.

Project E: Construction of a large mall, as well as restaurants and discos.

These projects are subject to the following conditions or criteria:

C1: Internal Rate of Return (IRR), in percentage, which must be superior to the return that the same amount of money could yield if placed in another investment. In this respect, there were estimated scores for each project, considering the results of the respective feasibility studies and financial statements, taking into account initial investments, bank loans and interest rates, working capital, potential patronage, etc., over a 30 years period.

C2: Net Present Value, in millions of Euros, considering the same period; scores were obtained as above.

C3: Payback period, which is the estimate of number of years for initial disbursements recovery, with information also from financial statements.

C4: Environmental disturbance. This is a serious matter due to environmentalists' movement against the project and regulations from the Ministry of the Environment. There is an agreement about scores, which values come from consultations on the impact of each alternative, surveys and polls, and discussions with environmental authorities. However, the Ministry of the Environment demands additional work to compensate for the damage that the undertaking – whatever it might be – will produce. This additional cost is already included in the budget.

Weights assigned to each of these criteria come from the DM and his team. Table 3.3 reflects these initial conditions, while Table 3.4 depicts normalized data.

Table 3.3 Decision matrix

		Criteria				
		Internal rate of return (%)	Net present value (millions of €)	Payback period (years)	Environmental disturbance (number)	
Criteria I.D.		C1	C2	C3	C4	
Criteria weights		0.29	0.26	0.18	0.27	Sum of rows
Action		Max	Max	Min	Min	
Construction of hotel and casino	A	**7.5**	**8.09**	**6.5**	**7**	29.09
Construction of a cable car and a restaurant at the top of the mountain	B	**6.9**	**8.54**	**7.1**	**6**	28.54
Construction of a large sporting facility including a marina	C	**8.2**	**9.01**	**6.6**	**6**	29.81
Construction of a cinema complex	D	**7**	**6.9**	**9**	**8.5**	31.4
Construction of mall	E	**8.2**	**7.4**	**5**	**8.5**	29.1

Table 3.4 Decision matrix normalized and weighted

Criteria I.D.		C1	C2	C3	C4
Criteria weights		0.29	0.26	0.18	0.27
Action		Max	Max	Min	Min
Construction of hotel and casino	A	**0.075**	**0.072**	**0.040**	**0.065**
Construction of a cable car and a restaurant at the top of the mountain	B	**0.070**	**0.078**	**0.045**	**0.057**
Construction of a large sporting facility including a marina	C	**0.018**	**0.079**	**0.040**	**0.054**
Construction of a cinema complex	D	**0.065**	**0.087**	**0.052**	**0.073**
Construction of mall	E	**0.082**	**0.066**	**0.031**	**0.07**

Comparisons to Build the Concordance Matrix

The process begins by comparing a pair of alternatives regarding a criterion. If the values of scores satisfy the action, then this pair receives a score equal to the corresponding criterion weight. For instance, let us consider pair A and B and criterion C1. The values for both scores, are 0.075 for A and 0.070 for B, which indicates that A>B. Since this criterion calls for maximization, this difference satisfies the action, and then this pair gets the **0.29** (this criterion weight) value. Consider now criterion C2 that calls for maximization, where A=0.072 and B=0.078, or A<B, therefore the action is not satisfied and a '0' placed. Similarly, consider the same pair for criterion C3 where A=0.040 and B=0.045, or A<B, then, because this criterion calls for a minimization, this difference does satisfy the criterion and the corresponding criterion weight of **0.18** is assigned to this pair. For criterion C4 that calls for a minimization A=0.065 and B=0.057, that is A>B and the action is not satisfied and a '0' is placed. If both scores are the same, just take half of the weight for the corresponding criterion. Table 3.5 shows this outcome for the first pair A and B, as 0.29+0.18=0.47.

Table 3.5 Calculation example

Criteria I.D.		C1	C2	C3	C4	
Criteria weights		0.29	0.26	0.18	0.27	
Action		Max	Max	Min	Min	Result for this pair
Construction of hotel and casino	A/B	**0.29**	**0**	**0.18**	**0**	**<u>0.47</u>**

Repeating this analysis for pairs A-C, A-D, A-E, B-C, B-D, B-E, C-D, C-E, and D-E, produces the results shown in Table 3.6, which is a square matrix with alternatives in columns and rows.

Table 3.6 Concordance index matrix

	A	B	C	D	E
A	–	**<u>0.47</u>**	**0.38**	**0.73**	**0.53**
B		–	**0.29**	**1**	**0.53**
C			–	**0.71**	**0.53**
D				–	**0.135**
E					–

Comparisons to Build the Discordance Matrix

A similar procedure as explained for the concordance matrix is now required, similar but not equal. Again, a pair of alternatives is compared regarding each criterion, but now we are not using the differences to assign a criterion weight but the differences in scores. Considering the same actions as before for each pair and each criterion, the absolute difference between both scores is computed. That is, for A and B pair and criterion C1 which calls for maximization and, since it satisfies the requirement, the absolute difference between scores is computed, that is $|0.075-0.070|$ (Table 3.4). Since for the second criterion the action is not satisfied, no value is computed (0). For C3 it will be $|0.040-0.045|$. For C4 it will be '0' since the action is not satisfied. The last column in Table 3.7 registers the maximum difference value, that is, 0.005. The same procedure is followed for all other pairs of alternatives.

Table 3.7 Calculation example according to action

Criteria I.D.		C1	C2	C3	C4	
Criteria weights		0.29	0.26	0.18	0.27	Result for this pair
Action		Max	Max	Min	Min	Maximum value of differences
Construction of hotel and casino	A/B	$|0.075-0.070|$	0	$|0.040-0.045|$	0	0.005

Now, it is also necessary to compute, for each pair and each criterion, the maximum value of the differences along each criterion (Table 3.8).

Table 3.8 Calculation example for the whole alternative

Criteria I.D.		C1	C2	C3	C4	
Criteria weights		0.29	0.26	0.18	0.27	Result for this pair
Action		Min	Min	Max	Max	Maximum value of differences
Construction of hotel and casino	A/B	$|0.075-0.070|$	$|0.072-0.078|$	$|0.040-0.045|$	$|0.065-0.057|$	0.008

Finally, a ratio is computed for each alternative considering the values found. That is 0.005/0.008 = 0.625, and then placed at the intersection of project A with project B in Table 3.9, which constitutes the Discordance index matrix. Making the computations for all the pairs, this table shows the values for this matrix.

Table 3.9 Discordance index matrix

	A	B	C	D	E
A	–	0.625	1	1	1
B		–	1	1	1
C			–	0.468	0.391
D				–	0.029
E					–

Now that we have both concordance and discordance matrices, it is time to introduce thresholds. Assume that the DM has assigned a minimum threshold value of c = 0.15 for the concordance matrix. Consequently, all values below this threshold are null, which is the case in pair D and E (Table 3.6). When this comparison is complete, the Concordance Dominance matrix is built (Table 3.10).

Table 3.10 Concordance dominance matrix

	A	B	C	D	E
A		1	1	1	1
B				1	1
C				1	1
D					
E					

The same is done with the Discordance index matrix but now using a maximum threshold of d = 1. There is no value larger than this threshold and Discordance Dominance matrix (Table 3.11) can be built.

Table 3.11 Discordance dominance matrix

	A	B	C	D	E
A		1	1	1	1
B			1	1	1
C				1	1
D					1
E					

The next step is a comparison of both dominance matrices, that is Tables 3.10 and 3.11. If a certain alternative agrees in outranking in both matrices, a 1 is considered. Thus, for alternative A for instance we found that in Table 3.10 it outranks B, C, D and E, and the same happens in the discordance matrix in Table 3.11.

There are also coincidences for pairs B and E, C and D, and C and E, but there are no coincidences for D and E. These coincidences allow construction of the Aggregated Dominance matrix, Table 3.12.

Table 3.12 Aggregated dominance matrix

	A	B	C	D	E
A		1	1	1	1
B				1	1
C				1	1
D					
E					

Note that alternative A outranks all the others, and consequently is the best. An arrow graph where an arrow indicates dominance from one alternative over the others allows better visualization, Fig. 3.2.

Node A corresponds to the nucleus or kernel and is the solution as the best alternative.

3.2.2.2 Comments on This Model

Pros.

1. Allows using fuzzy analysis because thresholds of indifference and preference.
2. Accepts qualitative and quantitative criteria.

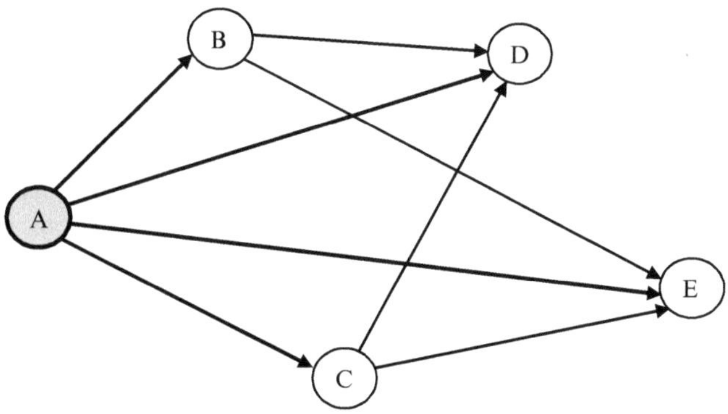

Fig. 3.2 Graph depicting values from matrix 3.12

3. Has the advantage of its multidimensional nature.
4. This method has a large advantage over others considering the two concordance and discordance matrices because the first one is related with criteria weights while the second is referred to scores, thus analyzing both effects independently and further combining them both. For a more technical discussion of the exact meaning of both matrices, see Van Delft and Nijkamp (1977).

Cons.

1. Doesn't have an axiomatic foundation.
2. It is difficult to understand, because of the principles used in determining the concordance and discordance matrices.
3. Thresholds can be calculated from these metrics, but are often established according to DM opinion which translates into subjectivity.

3.2.3 PROMETHEE-GAIA (Preference Ranking Organization Method for Enrichment Evaluations)

This method also belongs to the outranking models family and introduces concepts and parameters which pose some physical or economic interpretation easily understandable by the DM (Flament 1999). It makes extensive use of pseudo criteria and works by making comparisons between two alternatives and computing the difference, and then applying one of six *'transfer functions'*. (See also Guerrero Padilla et al. 2000).

An indispensable condition is that the different alternatives be comparable. For a given criterion 'i', and considering two alternatives 'a' and 'b', the difference between them will be equal to the difference of their scores, i.e.:

$$d_i(a,b) = v_i(a) - v_i(b).$$

Entering this value in any of the transfer functions, a value is found which is between 0 and 1, i.e.:

$$0 \leq P_i[d_i(a,b)] \leq 1.$$

This procedure is applied to each criterion and naturally it can be a maximization or a minimization criterion. Transfer functions are shown in Fig. 3.3 (Brans et al. 1986).

Where 'q' is the *indifference* threshold. This indifference level identifies which is the largest value of the difference by which 'a' is indifferent to 'b'. There is also a *preference* threshold 'p' which identifies the minimum value of the difference in which 'a' differentiates from 'b'. For instance if the difference between 'a' and 'b' is say 0.3 and the indifference level is 0.45, 'a' and 'b' are considered indifferent. Nevertheless, if this difference is larger, and this value is greater than the preference level, then 'a' > 'b'.

Consequently, in PROMETHEE it is necessary to use an evaluation table to establish weights for each criterion, select the transfer function for each criterion and finally set up thresholds. Notice that for a given value of the difference, there might be diverse values for the preference in accordance with the transfer function adopted. As mentioned, for each criterion, the model calls for choosing a preference function, and this selection depends on the characteristics of the criterion. Thus, for a criterion with a high uncertainty, it is convenient to use the Gaussian function; in another, where the difference is a direct relationship with quality for instance, it is convenient to use the linear function. Besides, observe that for small values of the difference, the function shows a value close to zero, that is there is indifference.

As hinted, in this method there is also a high subjectivity content, which manifests in the selection of the transfer function and in the fixing of the indifference 'q' and preference 'p' levels. There are several versions of PROMETHEE and each one built with a specific purpose. Very generally:

PROMETHEE I: Performs a partial ranking of alternatives since it considers only those where a strong preference exists and does not compare conflicting alternatives.

PROMETHEE II: Supplies a complete ranking of alternatives, which is based in a net result of positive flows (that is dominant alternatives), and negatives (i.e., dominated alternatives). This version utilizes sensitivity analysis to learn about stability of solutions when some parameters change, for example criteria weights.

PROMETHEE III: Works with assessed outranking relationships and also with problems involving fuzzy logic. Utilizes Integer Linear Programming[3] which uses as coefficients of the objective function data generated by PROMETHEE III.

PROMETHEE IV: Used when there are many alternatives.

PROMETHEE V: Applies Integer Linear Programming to select alternatives previously identified by PROMETHEE II and subject to a set of restrictions.

[3] Integer Linear Programming is a mathematical programming model with an additional condition demanding that variables in the result must be expressed as integers. It is a difficult problem to solve and still more restrictive than the LP model; normally uses the Gomory algorithm (Gomory 1958).

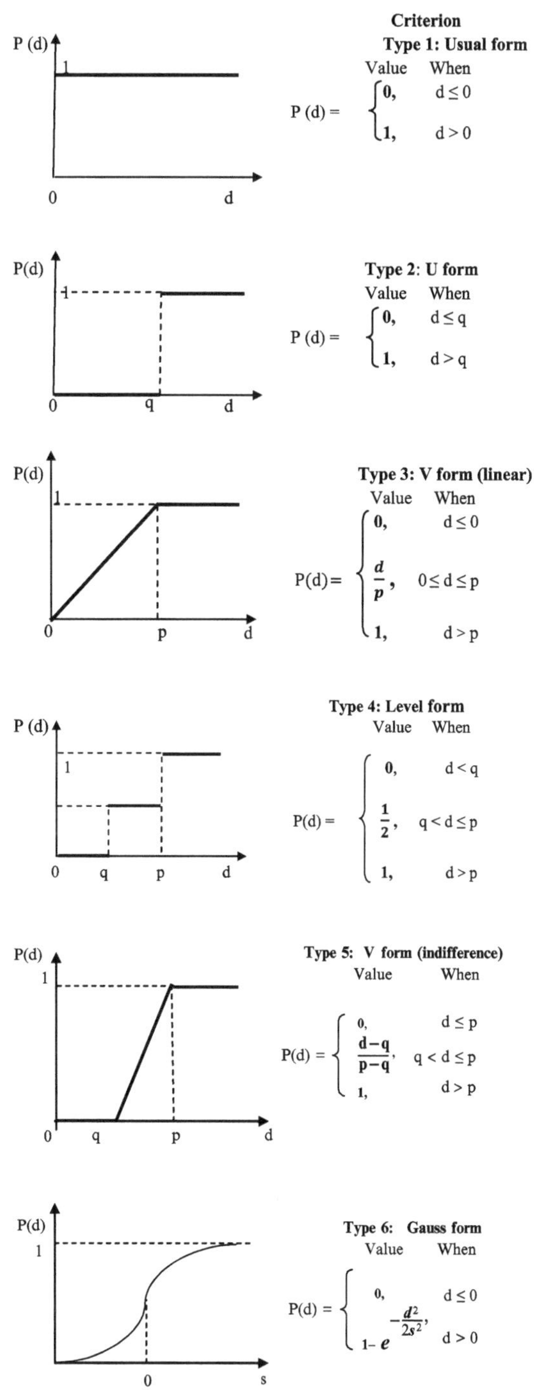

Fig. 3.3 Transfer functions

Computation sequence for the PROMETHEE model is as follows:

1. Prepare a decision matrix with alternatives in rows and criteria in columns.
2. Assign a weight to each criterion and select a transfer function for each one.
3. Establish preference thresholds ('p' and 'q') and indicate if it is maximization or a minimization criterion.
4. Start working in the first column with the two first alternatives (that is the two first rows) analyzing the difference between the values of two alternatives on that column. Compare this difference with the thresholds and apply corresponding formula for the selected transfer function. Then multiply this value by the weight assigned to that criterion.
5. When first row completed, add up all obtained values.
6. Build a square matrix or a preference indexes matrix, with the alternatives as rows and columns. Assign a zero at the intersection of an alternative in a column with the same alternative in a row, and put in each cell the value found in step 5. That is, in the row corresponding to alternative 'a' at the intersection with alternative 'b', place the value found in (5).
7. The procedure repeats for alternatives pairs.
8. When completed, add up values in each row and then calculate their average (remember to divide by the number of alternatives minus one, since always one of the values is zero). This average indicates the average positive flow, that is one that corresponds to the alternatives that generate it. Do the same for each column, which is the average negative flow because it corresponds to alternatives that receive it.
9. Since an alternative normally generates and receives flows, the difference of both assesses its value. The largest value of these differences signals the first alternative in the ranking and the balance of decreasing values allows for the ordering of the ranking of alternatives.
10. Sensitivity analysis for criteria takes place through variation of thresholds.

'*DecisionLab*'®[4] is the name of a dedicated software to resolve this problem. It also performs sensitivity analysis, by giving the information about how much it is possible to vary parameters without modifying the solution found.

3.2.3.1 Illustrative Example: Selection Between Two Alternative Routes (Solved by Hand)

This section is divided into two parts, (a) and (b).

(a) explains with an example how to proceed manually.
(b) develops a case study using the *DecisionLab*® software.

(a) Solving It Manually
Consider an elemental project with two alternatives routes and three restrictions (or criteria) as shown in Table 3.13. The alternatives refer to different routes

[4] Decision Lab software: http://visualdecision.com/dlab.htm

Table 3.13 Decision table

Data				
Action	Min	Min	Max	
Criteria →	Total distance (Km)	Total cost (millions of Euros)	Savings per year (millions of Euros)	
Criteria weights	0.22	0.55	0.23	
Type of transfer function selected for each criterion	Type 1 'Usual'	Type 2 'U'	Type 5 'Linear'	
Weight of each criterion	1	1	1	
Form of the transfer function		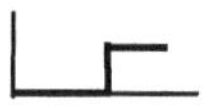		
Thresholds		q = 300	q = 10 p = 20	
Alternatives				
Route A	**1,820**	**1,389**	**36**	
Route B	**1,400**	**1,525**	**23**	
	Paired analysis of alternatives			
Comparing routes A and B	**0**	**0**	**0.07**	**0 + 0+0.07= 0.07**
Comparing routes B and A	**0.22**	**0**	**0**	**0.22 + 0 + 0 = 0.22**

between two points, and criteria are concerned with distances, costs and savings. The objective is to select the route that better satisfies the DM.

In reality it is not expected that the analyst will solve 'by hand' this problem doing all the calculations (it is cumbersome and very prone to errors), but it is believed to be important for him/her to understand what the software does, and this is the purpose of this example. The rules for working with PROMETHEE are easy to follow:

1. Prepare the decision table as shown in Table 3.13.
2. Specify the type of expected action for each criterion (Maximize or Minimize).
3. Specify the weight assigned to each criterion.
4. For each criterion, select the type of transfer function that you consider best fit the data. If there is uncertainty, one could choose the Gaussian.
5. Establish the thresholds 'q' and 'p' when applicable.

This is the data; now start with the analysis.

6. Consider a pair of alternatives regarding the first criterion (Total distance); primarily, check that this criterion calls for a minimization.
7. Check which of the two alternatives comply with this required *minimization* action. Evidently not A, because 1,820 > 1,400, or A > B, consequently *A does not dominate B*, and a '0' must be placed in cell A/B in column 1.

Examine now the second column and for the same pair. It calls again for a *minimization* action, but in this case this request is honoured by A since 1,389 < 1,525, or A < B. Therefore, *A dominates B.* Find the difference; in this case 1,525 – 1,389 = 136.

8. Check if this value is smaller than or greater than 'q'. Since q = 300, this difference is smaller, meaning that there is *indifference* and the function is worth '0'. That is, if it is true that A dominates B, the difference is less than the established threshold and then it is assumed that the value for A is '0'. Place this value in cell A/B column 2.
9. Go now to the third criterion in column 3, which calls for a *maximization* action. Since 36 > 23, or A > B, A honours the action. Find the difference; in this case 36–23 = 13. Compare this value with both thresholds. Since it is greater than 'q'(10) and smaller than 'p' (20) use the corresponding formula; (d–q)/(p–q) = (13–10)/(20–10) = 0.3. Multiply this value for the criterion weight (0.23), and place this result in the corresponding cell (0.30 × 0.23) = 0.07. Place this value in the cell A/B column 3.
10. Add up values in row for route A.
11. Proceed in the same way for the inverse pair B/A (second row).
12. Add up values in each row and the A/B = 0.07 and B/A = 0.22.
13. Now build Table 3.14, which is a square matrix formed by alternatives A and B.
14. According to the above values in A/B place 0.07 and in B/A place 0.22.
15. Add up rows which represent positive flow, which is the dominance of the alternative in that row, and columns, which is the dominated alternatives, that is the negative flow.
16. Deduct from the row corresponding to route A the column value for the *same alternative.* In this case it will be 0.07 – 0.22 = – 0.15. Do the same for route B. The reasons for doing this is that rows represent '*flows*' that originate there while columns represent '*sinks*' of flow, and for this reason the difference between these two flows for each alternative will be the net flow and represents its relative value. As can be appreciated, route B gets a much higher ranking than route A and it is consequently the preferable choice.

Naturally this was an elemental example; the same case but with more alternatives and criteria is solved in the next section using the dedicated software.

(b) Solving with Dedicated Software

Table 3.14 Paired matrix for alternatives comparison

	Route A	Route B	Sum of positive flow	Net flow (Positive – Negative)
Route A		0.07	0+0.07=0.07	0.07 – 0.22 = – 0.15
Route B	0. 22		0.22+0 = 0.22	0.22 – 0.07 = 0.15
Suma of negative flow	0+0.22=0.22	0.07+0=0.07		

3.2.3.2 Case Study: Route Selection for an Oil Pipeline Project (Solved Using DecisionLab® Software)

The Project

This project consists in building a pipeline to transport oil from the Black Sea area to Northern Italy for a distance of about 1,700 Km. There are three potential routes considered that differ in length, cost, topography, and difficulties. Along this distance, the pipeline traverses agriculture fields, swamps and forests, crosses rivers, passes a seismic zone and climbs mountains. The pipeline, mainly buried at an average depth of about 2 m, will run elevated (that is over the ground) in some areas, due to different reasons.

Criteria (in columns) are:

1. *Total distance*. Black Sea/Northern Italy. The proposed routes are:
   North Central: Crossing Romania, Hungary and Austria.
   Central: Pass through Romania, Bosnia-Herzegovina, Croatia and Austria.
   South Central: Through Romania, Serbia, Croatia, and Austria.
   These different routes constitute three different alternatives to evaluate. Each route is different, from the geographic, geological and political points of view and these factors are considered in this study.
2. *Cost*. Naturally, it has paramount importance, and it varies not only with the distance but mainly with the nature and magnitude of geographical accidents such as crossing rivers, swamps, agricultural land, forest, etc.
3. *People's opinion*. Very important concept which was evaluated by means of surveys and polls, to find how comfortable people are with the pipeline running near their cities and villages.
4. *Vulnerability*. Refers to how exposed the pipeline will be; that is how it can absorb certain risks especially those related with very cold weather, flooding, accessibility for repairs, response time from maintenance crews to arrive at a site with problems, etc.
5. *Savings*. Each route has a cost based not only on the construction activities, but also – very important – in a greater or lesser degree on the construction of facilities to move the oil. The latter involves pumping, control stations and service roads, and therefore there are different savings according to the length of the pipeline, the steepness in some sectors, the type of soil to excavate, etc.
6. *Agricultural land*. It is a very sensitive issue when the pipeline crosses agricultural land, and for several reasons this impact has to be kept at a minimum. Consequently, people's reaction to invasion of their lands and partial destruction of their method of livelihood, albeit with some compensation, must be evaluated. True, most of the pipeline goes underground, but even in that case, it is necessary to put in pumping and checking stations, as well as service roads that could use agricultural land.
7. *Elevation of pipeline*. In some areas, it is not convenient to have the pipe running underground because it is too expensive (such as trenching into solid rock),

therefore, it is sometimes convenient to lay the pipe over the ground albeit not touching it but elevated. Of course, there are also expenses for this activity, which besides increases the vulnerability of the system due to its visibility.

8. Sometimes there are *swamps to cross* and this is a very delicate subject for environmental reasons, consequently it is advisable to reduce crossing as much as possible.
9. *Earthquake risk.* All three alternatives face this risk. Unfortunately the three alternatives go into high risk areas, especially in Romania with moderate hazard in Croatia and Slovenia, and this must be considered.
10. *Sabotage risk.* The pipeline goes through highly sensitive political areas especially in the Balkans region, and this is something to take into account.

Criteria Weights

Assigned based on personal analysis, and shown in Table 3.15, which depicts all data.

Table 3.16 shows the same data in Decision Lab format, and Fig. 3.4 depicts the way in PROMETHEE 2, in which the result of the computation is displayed. It indicates the final result as the net flow of each alternative and also the ranking. The results and ranking are:

North Central = 0.11
South Central = 0.04
Central = −0.15

There are other outputs from this software albeit not shown here, such as the one called *'Walking Weights'*, which can be used for sensitivity analysis. It has a window that displays in the bottom the weights established by the DM, and it allows selecting any of the criteria and changing the weights. However, remember that changing say by 25% the weight in one criterion does not mean that the balance of criteria will change as well by the same 25%, and this feature shows it, indicating how this change in a certain criterion affects others differently.

Assume for instance that the DM wants to know the effect that a change in *'Total cost'* will have in the solution, and say that he/she wants to analyze the effect of about a 30% change in the weight, to bring it to 29% instead of 22% as it was before (shaded in Table 3.16). This is done using a ruler at the bottom of the window. The new values will show in horizontal solid bars. Then he/she can replace the old values for these new ones by going to the *'weight'* row in the decision table and making the substitution (shaded boxes in Table 3.17). Note how weights have changed in three criteria.

Running again the software, there will be a new result (Fig. 3.5), showing that with these new values the solution has changed to:

North Central = 0.17
South Central = −0.05
Central = −0.12

Table 3.15 Decision matrix for oil pipeline project

Alternatives routes →		**North central**	**South central**	**Central**			Thresholds	
Criteria I.D. and weights (%)		Alternatives scores			Action	Transfer function	**q**	**p**
Total distance	8	**1,820**	**1,549**	**1,612**	Min	Type 1		
Total cost	22	**1,389**	**1,719**	**1,418**	Min	Type 3		**40**
People's opinion	15	**40**	**50**	**30**	Max	Type 6	s=5%	
Vulnerability	15	**80**	**40**	**40**	Min	Type 6	s=6%	
Savings	7	**34**	**24**	**31**	Max	Type 5	**10**	**20**
Agriculture land	5	**27**	**84**	**23**	Min	Type 4	**11**	**27**
Pipe elevated	10	**27**	**45**	**0**	Max	Type 1		
Swamps crossed	6	**19**	**5**	**25**	Min	Type 6	s=3%	
Earthquake risk	7	**5**	**11**	**7**	Min	Type 6	s=5%	
Sabotage risk	5	**5**	**9**	**6**	Min	Type 6	s=6%	

Table 3.16 Decision matrix

Criteria →	Total distance	Total cost	People's opinion	Vulnerability	Savings	Agricultural land	Elevation of pipeline	Swamps crossing	Earthquake risk	Sabotage risk
Required action	Minimize	Minimize	Maximize	Minimize	Maximize	Minimize	Maximize	Minimize	Maximize	Minimize
Data										
Criteria weights	8	22	15	15	7	5	10	6	7	5
Preference function	Usual	V-Shape	Gaussian	Gaussian	Linear	Level	Usual	Gaussian	Gaussian	Gaussian
Indifference threshold					10	11				
Preference threshold		40			20	27				
Gaussian threshold (%)			5	6				3	5	6
Average performance	1,660	1,509	40	53	30	45	24	16	8	7
Standard deviation	142	173	10	23	5	34	23	10	3	2
Unit	km	Millions of €	%	%	Millions of €	km^2	ha	%	%	%
Alternatives										
North Central	1,820	1,389	40	80	34	27	27	19	5	5
South Central	1,549	1,719	50	40	24	84	45	5	11	9
Central	1,612	1,418	30	40	31	23	0	25	7	6

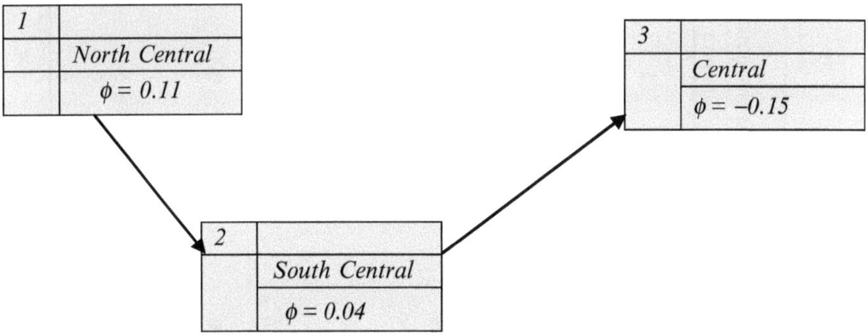

Fig. 3.4 Format in which Decision Lab presents the result

Therefore, the 30% increase in the weight for *'Total costs'* produces an increase in the already selected route, a decrease in the second one, and an increase in the third. Why?

Because in checking the decision table in Table 3.16 it is verified that North Central has the minimum cost (1,389) while South Central has the highest (1,719), and it is then obvious that an increase in costs weight will act favourably upon the selection with lesser costs. For the same token alternative 3 will increase slightly.

Doing this analysis for all the criteria, it is then possible to learn about their effects on the solution. It could very well be that the solution does not change even if there are certain changes in some criteria, and this is important, because it indicates on which criteria one must exert especial care or perform a more thorough scrutiny or research.

Another useful feature is the *'stability intervals'*, that when activated, shows between which limits a criterion can oscillate without changing the PROMETHEE 2 ranking.

GAIA Plane

This is another Decision Lab feature. The acronym stands for *'Geometrical Analysis for Interactive Assistance'* and that is exactly what it does. It assists the DM through a graphical representation of the decision problem. It is a useful device since it can indicate conflicting criteria. Combining the possible change of criteria weights in *'Walking Weights'*, with the GAIA plane, the analyst is able to watch interactively how the selection preferences change when changing the criteria weights, and visually appreciate in a graphical manner how one opposes the others.

3.2.3.3 Comments on This Model

Pros.

1. It is a logical and rational model, which everybody can understand and use.
2. It permits an easy and direct comparison of a pair of alternatives regarding one criterion.

Table 3.17 Decision matrix with different weights

CRITERIA →	Total distance	Total cost	People's opinion	Vulnerability	Savings	Agricultural land	Elevation of pipeline	Swamps crossing	Earthquake risk	Sabotage risk
Required action	Minimize	Minimize	Maximize	Minimize	Maximize	Minimize	Maximize	Minimize	Maximize	Minimize
Data										
Criteria weights	8	29	14	14	7	5	10	2	7	5
Preference function	Usual	V-Shape	Gaussian	Gaussian	Linear	Level	Usual	Gaussian	Gaussian	Gaussian
Indifference threshold					10	11				
Preference threshold		40			20	27				
Gaussian threshold (%)			5	6				3	5	6
Average performance	1,660	1,509	40	53	30	45	24	16	8	7
Standard deviation	142	173	10	23	5	34	23	10	3	2
Unit	km	Millions of €	%	%	Millions of €	km^2	ha	%	%	%
Alternatives										
North Central	1,820	1,389	40	80	34	27	27	19	5	5
South Central	1,549	1,719	50	40	24	84	45	5	11	9
Central	1,612	1,418	30	40	31	23	0	25	7	6

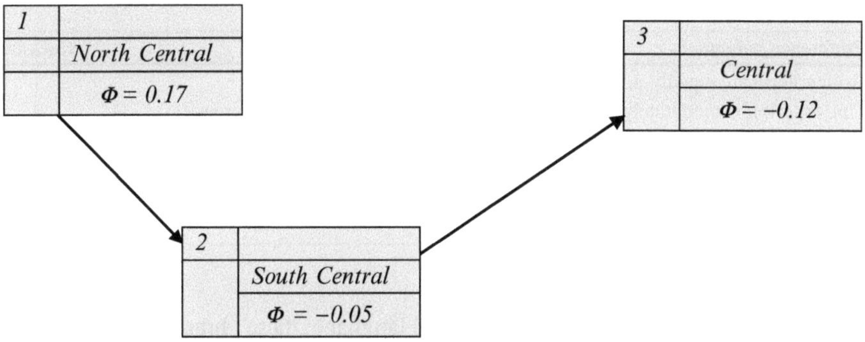

Fig. 3.5 Format in which Decision Lab presents the result when increasing weight in the 'Total cost' criterion

3. It is supported by good, effective and easy to use software with an innovative graphic system (GAIA) which allows visualizing different conditions especially related with changing criteria weight, which can greatly help the DM.
4. The transfer functions allow the analyst to consider the type of data available. Especially convenient is the Gaussian function in cases of uncertainty.

Cons.

1. There is subjectivity in establishing the criteria weights.
2. There is subjectivity in establishing the parameters 'q' and 'p'.
3. There is subjectivity in estimating which transfer function to use.

3.2.4 AHP (Analytic Hierarchy Process)

This model works by establishing preferences. In its first stage, it computes criteria weights. It starts by making paired comparisons between criteria, and builds a square matrix from which the eigenvector[5] is calculated, which is then utilized as a weight vector for criteria. In many applications, the AHP method stops here, however, the methodology also allows in a second stage for the determination of alternatives ranking, when alternatives are compared with each other regarding a specific criterion. This procedure is repeated considering all criteria, and the weighted summation of these values indicates the dominant alternatives, and hence the ranking.

This method is very popular in many different applications, some of them not conventional, such as for instance, in health care waste management (Brent et al. 2007) (See also Ghazinoory et al. 2008). One of the reasons for its diffusion and favouritism with practitioners is its simplicity and the fact that it establishes relationships between criteria and between alternatives according to the DM preferences,

[5] Eigenvector: German word meaning *'self vector'*. It is a property of matrices.

Table 3.18 Preferences scale

Preference judgement	Ranking
Both criteria are equally important or preferred	1
One criterion is moderately more important than the others (weak preference)	3
One criterion is strongly more important than the others	5
One criterion is very strongly more important than the others	7
One criterion is extremely more important than the others	9
Intermediate values	2-4-6-8

who can express them in phrases, or ordinal language; these phrases, using a certain scale, give way to cardinal numbers.

However, the method, called Analytic Hierarchy Process (AHP), has received some adverse criticism from the technical point of view because, amongst other aspects, it lacks a mathematical foundation for the scale used to convert ordinal concepts into cardinal values. Its author perhaps motivated by this criticism or just to improve the method, has developed an expanded version of the model, albeit based on the same principles, which is called Analytic Network Process (ANP). This is without a doubt more realistic, but also more laborious due to the larger number of comparisons required.

The model starts by identifying the criteria, which will be used to evaluate the different alternatives, as in all methods, and then, proceeds as indicated by the following steps:

1. It starts by building a square matrix using the same criteria in columns and in rows.
2. Then, it makes a pair comparison of criteria using preferences, regarding the objective as a reference, and assigning a valuation measure between 1 and 9, as per Table 3.18. For instance, in comparing criteria A and B:

   If the DM prefers criteria A over B, say, with a weak preference, he/she uses Table 3.18 to convert that preference judgement into a cardinal number. The resulting '3' value means that A is three times more significant than B, regarding the objective. Consequently, the model assumes that B is exactly the opposite, and gets 1/3. The same procedure applies to compare all pairs of criteria.
3. Now, it finds the eigenvector. (The software does it, if not, the analyst can follow the simple procedure indicated ahead).
4. To weight alternatives a similar procedure applies. Now, it makes pair comparison of alternatives with respect to one criterion. The result is a new reciprocal square matrix for each criterion, with its corresponding eigenvector. When this procedure is repeated for all criteria, there will be a value or weight for each alternative and for each criterion.
5. Afterwards, it multiplies the value of each alternative by the weight of the corresponding criterion.
6. Finally, it adds up all values for an alternative. The final figure indicates the importance of each one.

3.2.4.1 Illustrative Example: A Travel Dilemma (Solved by Hand)

As an example, assume for instance that somebody needs to make an assessment about different modes of transportation, namely by car, train or bus between City 1 and City 2. To evaluate them, one uses four different criteria or parameters: Travel time, Travel cost, Comfort and Safety. The objective is to determine the most convenient way to travel.

As per the steps indicated above:

Step 1 and Step 2

Prepare the square matrix in Table 3.19 to compare criteria using preference values from Table 3.18.

Table 3.19 Inverse matrix to compare criteria

	A Travel time	B Travel cost	C Comfort	D Safety
A Travel time	1	0.5	1	1
B Travel cost	2	1	5	0.33
C Comfort	1	0.2	1	2
D Safety	1	3	0.5	1

Start by comparing criterion '*travel time*' and criterion '*travel cost*'. The DM (the traveller) is then asked (or asks him/herself), the following question: 'Considering the most convenient way to travel, which is more important in this trip, travel time or travel cost?' The traveller could answer: "*Although for me both criteria have similar relevance I believe that travel cost is a little more important*". Since there is not a strong preference of travel cost over travel time, a 2 (intermediate value from Table 3.18) is assigned to the intersection of row B (travel cost) with column A (travel time). This means that the traveller thinks that travel cost is two times more important than travel time or that, according to this criterion, travel cost dominates travel time. Conversely, in the intersection of row A with column B the inverse value is assigned, i.e. 0.5, or that travel time is dominated by travel cost.

When similar questions are formulated comparing the criterion travel time with the criteria comfort and safety, assume that the traveller has shown indifference and because of that, '1s' are allocated to them both. The same procedure applies in comparing travel cost with comfort and safety, and finally comfort with safety.

Step 3

To compute the criteria weights using the values thus obtained, we need to use linear algebra and compute the eigenvector. However, there is an approximate procedure, which is less complicated. Simply multiply all the quantities in one row, add these products up, and extract the root corresponding to the number of criteria, in this case 4.

As an example, for row A in Table 3.20:

Table 3.20 Weights computation by the approximate method

Criteria			Criteria relative weights
Column 1	Column 2	Column 3	Column 4
A Travel time	$1 \times 0.5 \times 1 \times 1 = 0.5$	0.84	0.21
B Travel cost	$2 \times 1 \times 5 \times 0.33 = 3.3$	1.35	0.33
C Comfort	$1 \times 0.2 \times 1 \times 2 = 0.4$	0.79	0.19
D Safety	$1 \times 3 \times 0.5 \times 1 = 1.5$	1.11	0.27
Total		4.09	

To find the relative weight of each criterion, divide each one in column 3 by this total. Notice that criterion B has the largest value.

Step 4

If the method is used to select alternatives, and considering that these are three, a similar procedure is followed, determining now a reciprocal square matrix for all alternatives and for each criterion; the comparison between pairs of alternatives is done considering sequentially one criterion at a time, as shown in Table 3.21 for criterion *'travel time'*.

Table 3.21 Alternatives weight for criterion 'travel time'

Alternatives	1 Car	2 Train	3 Bus	$\prod_{i=1}^{i=3}$		Alternatives weight
1 Car	1	0.8	1.12	0.896	0.964	0.303
2 Train	1.25	1	③	3.750	1.554	0.488
3 Bus	0.893	0.333	1	0.298	0.668	0.210
Total:					3.186	

In this example of travelling between City 1 and City 2 the formulated question could be *"From the point of view of travel time, how do you evaluate each alternative?"* For instance the DM could answer, *"I prefer the train to the bus"*, and then, because it is rather a weak preference and not a strong one, a 3 is placed in row 2 *(train)* and column 3 *(bus)*, according to the preference scale in Table 3.18. Notice that train is also mildly preferred to travelling by car, and this preference gets the 1.25 value, while preference of car over train gets its inverse, 0.8. Repeat this analysis for each pair of alternatives to complete all comparisons for this criterion.

A similar analysis applies to criteria *'Travel cost'*, *'Comfort'* and *'Safety'*. Values obtained for alternatives *'Car'*, *'Train'* and *'Bus'* regarding each criterion are shown in Table 3.22 for them all.

Table 3.22 Getting alternatives values for all criteria

	Alternatives		
Criteria	1 Car	2 Train	3 Bus
A Travel time	0.303	0.488	0.210
B Travel cost	0.279	0.197	0.524
C Comfort	0.262	0.444	0.295
D Safety	0.203	0.207	0.311

This matrix shows in each cell the value of the geometric mean of an alternative when it is compared with all the other alternatives and using a certain criterion. These values come from the weight column in each table. Thus, value 0.197 in cell *'Travel cost/Train'*, is the geometric mean of all values obtained when alternative *'train'* is compared with alternatives *'car'* and *'bus'* and using for comparison criterion 'travel cost'.

Step 5

Finally, apply criteria weights to these values to get results depicted in Table 3.23.

Table 3.23 Weighted values for alternatives

		Alternatives		
		1 Car	2 Train	3 Bus
Criteria	*Criteria weights*	*Alternatives weights*		
A Travel time	0.20	**0.061**	**0.097**	**0.042**
B Travel cost	0.33	**0.092**	**0.065**	**0.173**
C Comfort	0.19	**0.05**	**0.084**	**0.056**
D Safety	0.28	**0.057**	**0.058**	**0.087**
	Sum:	**0.260**	**0.304**	**0.358**

Step 6

Adding up values for each alternative indicates that the most preferable is alternative 3, which is travelling by bus.

This was only an explanatory example to illustrate how the method works; now a case study will be proposed and solved with dedicated software.

3.2.4.2 Case Study: Location Analysis for an Agriculture and Farming Machine Manufacturer (Solved Using AHP 'Super Decisions' Software)

Background Information

A large North American company produces a line of machinery for agriculture and farming, and wants to be present in the European market, considering the large amount of its equipment manufactured in North America and sold in Europe.

Naturally, the equipment being made in Europe will have a lower price than the same equipment produced in the USA because of the elimination of costly transportation from North America. The company has already made a preliminary analysis about different prospective sites and come down to three potential locations, namely Madrid, Paris and Frankfurt. This analysis considers seven criteria as follows:

(A) Regional demand, mainly in Central Europe.
(B) Subsidies offered by the above-mentioned cities if the factory, employing more than 550 workers, is located in their jurisdictions. These incentives relate with a rebate of taxes over a 10-year period and a monthly contribution of 150 Euros per worker during the two first years.
(C) Access to harbours: This is important for the company to be able to ship the equipment to South and Central America, the Middle East and Africa
(D) Distribution network: It refers to the facilities to ship equipment by road to different places in Europe.
(E) Available land: The firm is thinking of a plot of land of about 50 ha nearby any of the three potential sites.
(F) Qualified suppliers and with experience in this type of manufacturing process, because the plant will work on the *'just-in-time'*[6] principle.
(G) Manpower. It is important to have a pool of experienced workers in metalworking, engines assembly, and metal stamping.

Objective: To install the manufacturing plant in a location that offers the best advantages from the economic point of view.

Using SuperDecisions®[7] software and placing the problem in a graphic perspective.

In order to select the most adequate location they want to use the AHP method and work with the dedicated software called *'Super Decisions'*.[8] Figure 3.6 shows the hierarchical levels typical of this model. That is, criteria are compared one with each other regarding the objective, while alternatives are compared one with each other according to each criterion.

Determining Criteria Weights

Compare criteria as explained in Table 3.19 in Sect. 3.2.4.1, and build Table 3.24. Only half of the matrix is depicted since the other half (down the main diagonal) is the

[6] Manufacturing strategy: Consists in reducing as much as possible the material, parts, components etc., in process, especially in assembly lines.

[7] Software for decision-making; http://www.superdecisions.com/.

[8] In reality, this software is used to solve a much larger type of problems with many more relationships between its components. This is the so-called *'Analytical Network Process'*.

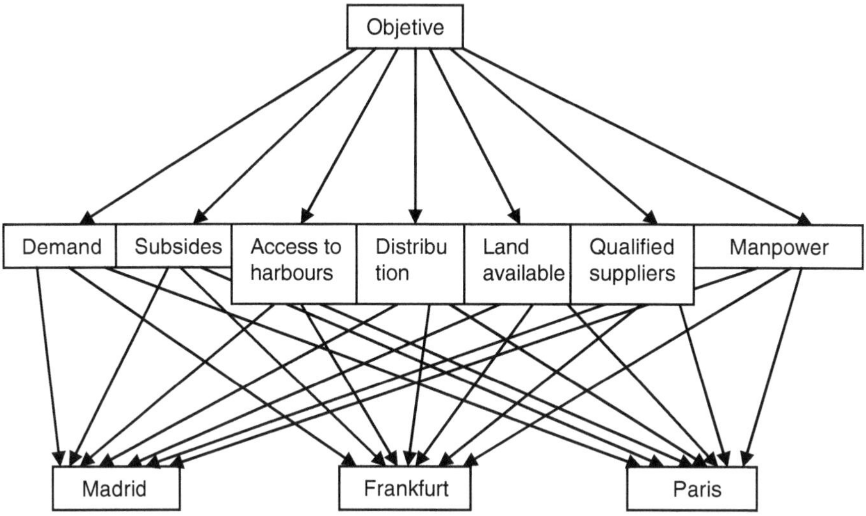

Fig. 3.6 Hierarchy for this case

Table 3.24 Comparison between criteria respective to objective

	Regional demand	Subsides	Access to harbours	Distribution network	Available land	Qualified suppliers	Manpower
Regional demand		2	6	4	4	8	8
Subsides			3	6	2	2	6
Access to harbours				2	2	6	8
Distribution network					1	1	2
Availableland						2	6
Qualified suppliers							3
Manpower							1

inverse of the other and the software recognizes it). As explained, the cardinal values depicted in Table 3.24 come from applying the *'Preferences scale'* (Table 3.18).

These values are inputted into the software through some sort of numerical pad (not shown here, but sketched in Fig. 3.7), which has a pair of criteria in each row, and two series of numbers from 1 to 9 in two sectors. The left sector corresponds to preferences when the criterion in the left is larger than the criterion at the right. The right sector belongs to preferences when the criterion at the right has a higher preference than the criterion at the left. Value 1 is shared by both sectors, that is when there is *'indifference'*, or when both have *'equality importance'*.

In this problem, the first row in the pad corresponds to pair *'Regional demand'*, on the left and *'Subsidies'* on its right. In the second row, it shows *'Demand'* on its left and *'Access to harbours'* on the right. In the third row, it is the pair *'Demand'* on the left and *'Distribution network'* on its right, and so on. Then, we have the layout sketched in Fig. 3.7.

A.	Demand	9	8	7	6	5	4	3	**2**	1	2	3	4	5	6	7	8	9	**B.** Subsidies
A.	Demand	9	8	7	**6**	5	4	3	2	1	2	3	4	5	6	7	8	9	**C.** Access to harbours
A.	Demand	9	8	7	6	5	**4**	3	2	1	2	3	4	5	6	7	8	9	**D.** Distributionnetwork
A.	Demand	9	8	7	6	5	**4**	3	2	1	2	3	4	5	6	7	8	9	**E.** Available land
A.	Demand	9	**8**	7	6	5	4	3	2	1	2	3	4	5	6	7	8	9	**F.**Qualified suppliers
A.	Demand	9	**8**	7	6	5	4	3	2	1	2	3	4	5	6	7	8	9	**G.**Manpower

Fig. 3.7 Sketch of super decisions numerical pad

Thus, it shows that criterion A (Regional demand) is 2 times more important than criterion B (Subsides), 6 times more important than criterion C (Access to harbours), and so on. If for instance it were found that criterion D *'Distribution network'* were say, 4 times more important than criterion B *'Subsidies'*, then the numerical pad would look like this:

B. Subsidies 9 8 7 6 5 4 3 2 1 2 3 **4** 5 6 7 8 9 **D**. Distribution network.

Once this input is complete, the software makes the calculation commented on in Sect. 3.2.4.1, and shows as a result the criteria weights as depicted in Fig. 3.8. Note at the top the comment on the inconsistency index. This is an AHP and ANP feature to make sure that there is consistency (transitivity) in the personal evaluations, in other words, it guarantees that transitivity is maintained (about the meaning for *'transitivity'* go to Sect. 3.2.4.3 *Cons.* 1).

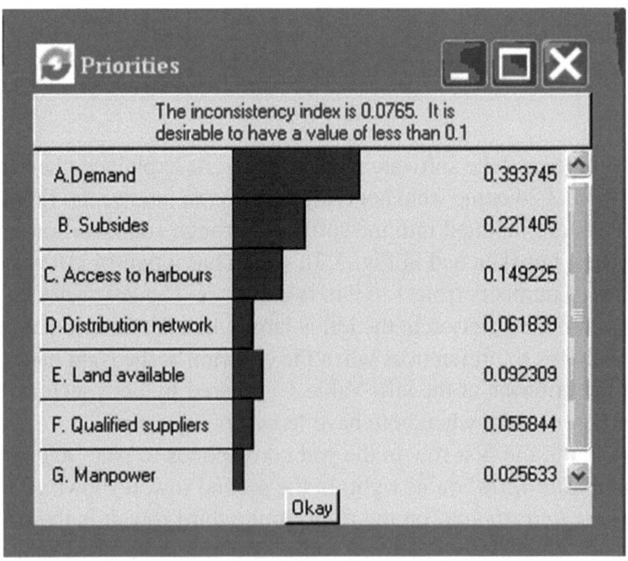

Fig. 3.8 Computer output weights calculated for each criterion. (Super decisions screen shots reproduced with kind permission of Thomas Saaty, developer of this program)

As expected the most important criterion is *'Demand'*, with *'Subsidies'* in second place and *'Access to harbours'* in third place. Here finishes the determination of criteria weights, or first stage of the AHP method. The second stage is the determination of the alternatives weights or their rankings.

If the purpose is to select alternatives, such as in this case, then it is necessary to analyze each alternative on each criterion; we can start with criterion *'Regional Demand'* (or any other), and compare alternatives regarding this criterion. In this example, Table 3.25 shows that, according to the DM preferences, Madrid is twice more important than Paris, and four times more important than Frankfurt, and gets the highest weight. Table 3.25 completes the values obtained out of this comparison.

Table 3.25 Alternative criteria comparison for criterion 'Regional Demand' (A)

	Madrid	Paris	Frankfurt	Weight
Madrid		2	4	**0.558**
Paris			3	**0.122**
Frankfurt				**0.319**

Values in the last column of Table 3.25 are the weights, corresponding to the eigenvector, computed by *Super Decisions* as shown in Fig. 3.9.

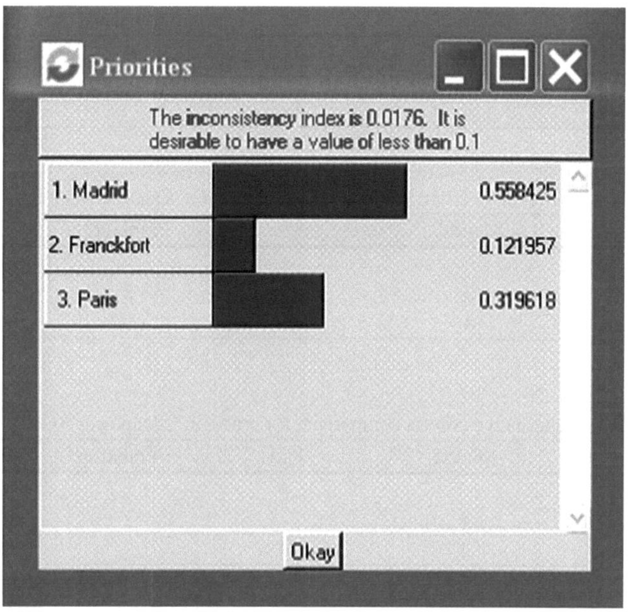

Fig. 3.9 Software provides values of eigenvectors. (Super decisions screen shots reproduced with kind permission of Thomas Saaty, developer of this program)

The procedure is repeated for each criterion as depicted in Tables 3.26–3.31.

The software, in a next step, multiplies each weight found for an alternative by the corresponding weight assigned to each criterion. It adds up these results for each alternative and delivers the result shown in Fig. 3.10.

Table 3.26 Alternative criteria comparison for criterion 'Subsidies' (B)

	Madrid	Paris	Frankfurt	Weight
Madrid				**0.124**
Paris	**3**		**2**	**0.517**
Frankfurt	**4**			**0.358**

Table 3.27 Alternative criteria comparison for criterion 'Access to harbour' (C)

	Madrid	Paris	Frankfurt	Weight
Madrid		**2**	**4**	**0.571**
Paris			**2**	**0.285**
Frankfurt				**0.142**

Table 3.28 Alternative criteria comparison for criterion 'Distribution network' (D)

	Madrid	Paris	Frankfurt	Weight
Madrid				**0.089**
Paris	**4**		**3**	**0.587**
Frankfurt	**6**			**0.323**

Table 3.29 Alternative criteria comparison for criterion 'Available land' (E)

	Madrid	Paris	Frankfurt	Weight
Madrid		**4**	**4**	**0.651**
Paris				**0.113**
Frankfurt		**3**		**0.235**

Table 3.30 Alternative criteria comparison for criterion 'Qualified suppliers' (F)

	Madrid	Paris	Frankfurt	Weight
Madrid				**0.164**
Paris	**3**		**2**	**0.139**
Frankfurt	**4**			**0.297**

Table 3.31 Alternative criteria comparison for criterion 'Manpower' (G)

	Madrid	Paris	Frankfurt	Weight
Madrid				**0.149**
Paris	**4**			**0.474**
Frankfurt	**2**	**1**		**0.376**

Then:

Madrid = 0.20
Paris = 0.18
Frankfurt = 0.11

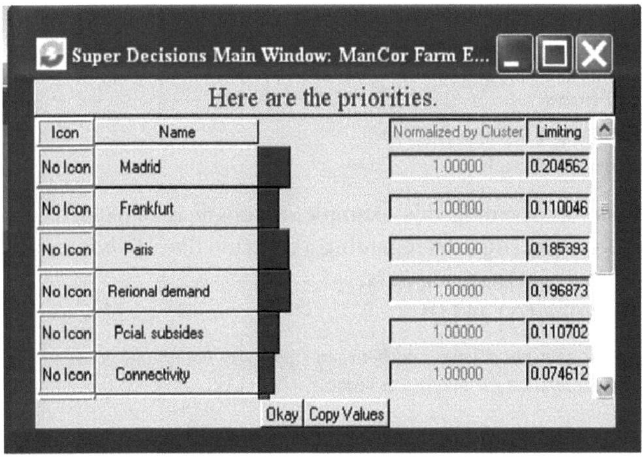

Fig. 3.10 Super decisions shows the final values for the three alternatives. (Super Decisions screen shots reproduced with kind permission of Thomas Saaty, developer of this program)

As can be seen the winner is Madrid as expected, because it has the largest alternative weight (0.558), for the criterion with the largest weight (0.394), and the same happens for criterion *'Access to harbours'*.

3.2.4.3 Comments on This Model

Pros:

1. It is by far the most popular model and used extensively in a variety of activities.
2. It is easy to use, and the availability of effective and uncomplicated software is the reason for this popularity.
3. Probably its best advantage is that it allows the DM to set up a project in a manner that many people are used to, that is expressing their preferences, and from this point of view this is a unique and valuable feature of this method.
4. It is very intuitive.
5. It is easy to understand even if the analyst does not grasp the mathematical meaning of the eigenvectors.
6. One of the advantages of the method (the Analytical Network Process, not the Analytical Hierarchy Process), is that it can contemplate series of effects, and this is important, because usually there is chain of effects provoked just for one of them but that travels as a ripple along a chain (the so-called domino effect).

Cons:

1. In this writer's opinion the model, although it could be mathematically correct in its application, does not necessarily reflect the reality, because it is based on the concept of *'mathematical transitivity'*, which could be not rational in a real life

situation. To illustrate this assertion assume for instance three alternatives, and based only in preferences, of places to spend vacation time:

(A) Stay at home.
(B) Vacation in a mountain area.
(C) Move to the beach.

Suppose that, concerning this example, a person is requested to voice his/her preferences on this situation regarding a criterion like *'To have a good time'*; it is then possible to get these answers:
When comparing (A) and (B):

Regarding to have a good time, and between staying at home and going to the mountain, I have a mild preference for staying at home.

And when comparing (B) and (C) the person could say:

Regarding having a good time, and between going to the mountain and spending time in a beach, I would perhaps prefer the mountain.

From here, according to AHP – because it assumes mathematical subjectivity – when comparing (A) and (C) i.e. '*staying at home'* (A) and '*going to the beach'* (C) the questioned person should say:

Regarding having a good time, and between staying at home and going to the beach, I strongly prefer staying at home,

Since, if A > B, and B > C, then A > C.

However, this might not be the answer because the person could say instead:

Regarding having a good time, and between 'staying at home' and 'going to the beach', I would rather prefer going to the beach.

Therefore, this consideration of transitivity can be a serious drawback because it introduces a bias into the model. Added to this shortcoming, there is also a numerical value expressing these preferences, producing results that may not be realistic. That is, if A is for instance moderately preferable to B, then the preference is worth a value of 3, in accordance with Table 3.18, and if B is 2 times preferable to C, then A is 6 times preferable to C. This does not look very reasonable because it is assuming that even if the person's preference were indeed staying at home, it is also saying that the person feels between strongly and very strongly that he/she prefers to stay at home, and of course, that might not be the case.
As a bottom line it is clear that necessarily AHP and ANP must have transitivity where two options are compared (A and C) using an intermediate B, which is not always verified in practice, and consequently the result may be biased.
Other researchers like Pomerol and Barba-Romero[9] (2000) share this opinion, and they state too that, when there is an indifference relationship between two

---

[9]This work provides a very interesting history on multicriteria decision and traces its origins to around 1785 with work of the Marquis de Caritat de Condorcet.

actions, this is not transitive. For instance, assume that we evaluate two alternatives for an airport through two criteria A and B. The first establishes the average quantity of flights for which a passenger must wait in the queue for no more than 12 min. Criterion B establishes the same but only for 3% of passengers.
When the DM is asked about which of these two criteria is more important he/she may say, *"They are indifferent for me"*, and probably with reason. However, what happens if in lieu of 12 min we are considering an hour or more, as is usual in international flights. The problem changes because we are talking about 97% of passengers waiting up to an hour, which is not precisely the same as waiting for 12 min, and in this case most probably the answer is not indifference.
2. The model works with much subjectivity. There are subjectivities in the preference scale for the criteria, not only because it is a subjective value but also in the scale by itself. There is subjectivity in applying the preferences scale in alternatives selection. Finally, the fact that two different persons can get different values for the same problem, casts a doubt about the reliability of this method. See also Dyer (1990).
3. Use of the eigenvector. Some researchers question the use of this technique. Bana e Costa and Vansnick (2001) after an extensive analysis say that they consider that AHP possesses a *"functional weakness which converts to problematic its use as a tool for decision-making"*.
4. The system is too time consuming. In a problem with say 50 criteria, there are thousands of comparisons to make. This can discourage the DM, and even with the best goodwill it is only human that perhaps hundreds of comparisons are made without too much thinking, and just to get rid of the task.

Sensitivity analysis

1. The feature *'Sensitivity'* displays a window with lines corresponding to different alternatives. In selecting any one of the criteria, it displays the field of validity of each alternative, and it is possible to combine several criteria.
2. There is another useful feature called *'Optimization'* which gives the lower and upper bounds between which a criterion can vary without changing the solution.

3.2.5 TOPSIS (Technique for Order Preference by Similarity to Ideal Situation)

3.2.5.1 Background Information on the Method

This is a simple and easy to understand method. It belongs to the family of methodologies that work with recognition of how far away a found solution is from the optimal solution; of course, the shorter the distance the better.

There are three steps as follows:

First step. Compute for a given problem its *ideal* solution. This is known since it is determined by the *best score* for each criterion with respect to maximization and minimization. That is, in a criterion that calls for maximization, the best score is the largest. If the criterion calls for minimization, the best score will be the smallest. In any case it is identified as vector A^+.

Compute now the *worst* solution. This is known since it is determined by the *worst score* of each criterion with respect to maximization and minimization. That is, in a criterion that calls for maximization, the worst score is the smallest. If the criterion calls for a minimization, the worst score is the largest. In any case it is identified as vector A^-.

In other words when actions of all criteria are optimized the result is the ideal solution, while when none of them are satisfied we are in presence of the worst solution.

Second step. For the first alternative, find the difference between its score in the first criterion and the best value for that criterion, in A^+, and square this difference; do it again, i.e. find the difference between the score of this first alternative and the second criterion and its best, in A^+, and square the difference. Repeat the process for all criteria. When completed, add up all these squared values and extract the square root of this sum; this is the Euclidean Metric, and it is the solution S_1^+ for the first alternative. Repeat the same procedure for all alternatives.

Repeat the same procedure but now considering worst values that is A^-.

As a final result there will be a matrix whose columns are alternatives and whose rows are the best $(R)^+$ and worst distances $(R)-$.

Third step. Compute a closeness index by using the following formula for each alternative:

$$\text{Closeness index }(\text{C.I.}) = \frac{(R)^-}{(R)^+ + (R)^-}$$

The alternative with the largest C.I. is the first to be calculated, followed by the others in decreasing C.I. order.

Figure 3.11 illustrates for a simpler case how the Euclidean Metric is used. In actual cases there are usually many alternatives and many criteria which make the problem multidimensional.

$$\begin{aligned}\text{Distance AB} &= \sqrt{(x_2 - x_1)^2 + (y_2 - y_1)^2} \\ &= \sqrt{(8-3)^2 + (7-4)^2} \\ &= 5.83.\end{aligned}$$

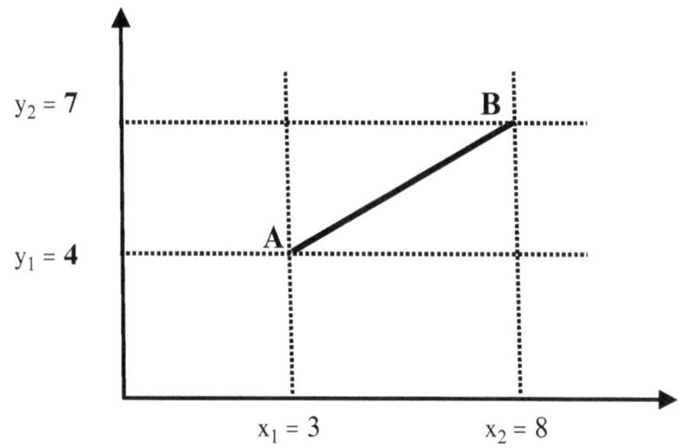

Fig. 3.11 Calculation of distance AB through Euclidean metric (not at scale)

3.2.5.2 Illustrative Example: Route Selection for a Transmission Line (Solved by Hand)

A company wants to erect an aluminium smelter in a seashore location, appropriate to receive bauxite[10] coming by sea. A hydro-power station especially built for this purpose at a distance of about 420 km from the future smelter, will generate the necessary electrical energy, which is almost a raw material in aluminium manufacturing. A transmission line is to be built linking the power station and the smelter, and because of the nature of the terrain in between, there are three alternative routes. The first one follows the Suquia River Valley, and is also the longest.

Another alternative is to build the line through the Trevel Mountains, which poses some challenges due to altitude and slopes. A third alternative route runs in about 30% of total distance through the Suquia River Valley and 70% through the Trevel Mountains. Both routes traverse beautiful scenic areas with many small cities and villages whose main activity is tourism; therefore both people's opinion and eco-system protection are very important characteristics to take into consideration. Table 3.32 depicts the decision table for this undertaking, as well as criteria weights.

Table 3.33 depicts all data with an added column, used for normalization, which shows the total value for each row; normalization is done by dividing each score in a row by that total value (Table 3.34).

Table 3.35 shows the weighted scores, thus, for the Suquia River Valley and total distance, the weighted score is now $0.353 \times 0.0\ 8 = 0.028$, and so on. Notice two additional columns labeled (A^+) and (A^-); these indicate best values.

For instance, for the first criterion *'Total distance',* (A^+) shows the best score (since it calls for a *minimum,* the *minimum* score is selected, which is then the ideal), while for another criterion such as *'People's opinion'*, the column shows the best score (because it calls for a *maximum,* the *maximum* score has been selected). This

[10] Bauxite is aluminium ore. A combination of iron and aluminium hydroxides mixed with other compounds such as oxides of Al, Fe, and other elements.

Table 3.32 Decision table for transmission line

Alternatives routes		Suquia river valley	Trevel mountains	Suquia valley/ mountains	
Criteria I.D. and weights (%)		Alternatives scores			Action
Total distance	8	**444**	**395**	**419**	Min
Total cost (Euros)	22	**1,389**	**1,719**	**1,418**	Min
People's opinion	15	**40**	**50**	**30**	Max
Slope	15	**80**	**40**	**40**	Min
Savings in transmission towers	7	**34**	**24**	**31**	Max
Crossing scenic area	5	**27**	**84**	**23**	Min
Ecosystem protection	10	**27**	**45**	**23**	Max
Swamps crossed	6	**19**	**5**	**25**	Min
Earthquake risk	7	**5**	**11**	**7**	Min
Forestry	5	**5**	**9**	**6**	Min

Table 3.33 Data matrix

	Suquia river valley	Trevel mountains	Suquia valley/ mountains	Action	Sum of rows
Total distance	**444**	**395**	**419**	Min	1,258
Total cost	**1,389**	**1,719**	**1,418**	Min	4,526
People's opinion	**40**	**60**	**30**	Max	130
Slope	**80**	**40**	**40**	Min	160
Savings in transmission towers	**34**	**24**	**30**	Max	88
Crossing scenic area	**27**	**84**	**23**	Min	134
Ecosystem protection	**27**	**67**	**23**	Max	117
Swamps crossed	**19**	**2**	**10**	Min	31
Earthquake risk	**4**	**11**	**7**	Min	22
Forestry	**5**	**10**	**7**	Min	22

Table 3.34 Normalization of data matrix

	Suquia river valley	Trevel mountains	Suquia valley/ mountains	Action
Total distance	**0.353**	**0.314**	**0.333**	Min
Total cost	**0.307**	**0.380**	**0.313**	Min
People's opinion	**0.308**	**0.462**	**0.231**	Max
Slope	**0.500**	**0.250**	**0.250**	Min
Savings in transmission towers	**0.386**	**0.273**	**0.341**	Max
Crossing scenic area	**0.201**	**0.627**	**0.172**	Min
Ecosystem protection	**0.231**	**0.573**	**0.197**	Max
Swamps crossed	**0.613**	**0.065**	**0.323**	Min
Earthquake risk	**0.182**	**0.500**	**0.318**	Min
Forestry	**0.227**	**0.455**	**0.318**	Min

Table 3.35 Weighted decision matrix

	Suquia river valley	Trevel mountains	Suquia valley/ mountains	Action	(A^+) Ideal solution	Action	(A^-) Worst solution
Total distance	0.028	0.025	0.027	Min	0.025	Max	0.028
Total cost	0.068	0.084	0.069	Min	0.068	Max	0.084
People's opinion	0.046	0.069	0.035	Max	0.069	Min	0.035
Slope	0.075	0.038	0.038	Min	0.038	Max	0.075
Savings in transmission towers	0.027	0.019	0.024	Max	0.027	Min	0.019
Crossing scenic area	0.010	0.031	0.009	Min	0.009	Max	0.031
Ecosystem protection	0.023	0.057	0.020	Max	0.006	Min	0.020
Swamps crossed	0.037	0.004	0.019	Min	0.004	Max	0.037
Earthquake risk	0.013	0.035	0.022	Min	0.013	Max	0.035
Forestry	0.011	0.023	0.016	Min	0.011	Max	0.023

column is then the ideal solution since this vector complies with all demands as their optimum values.

On the contrary, column (A^-) shows the worst situation. For instance, since the first criterion calls for a *minimum*, the opposite is selected, i.e. the *maximum*, which is the worst, and the same for other criteria.

Now, for each alternative the method computes the difference between its actual score on a certain criterion, and the best value for that criterion and this is done for all alternatives and all criteria. This is repeated for the worst scenario.

What we are trying to determine is the value of the *distance* of each alternative in a multidimensional scenario. As mentioned, a Euclidean Metric is applied, as the square root of the sum of the squared differences between scores and the best ones. Table 3.36 shows the final result $(R)^+$, as well as $(R)^-$ of this calculation for all alternatives.

As an example, for the Suquia River Valley alternative the depicted value comes from the following formula:

$$(R)^+ = \sqrt{(0.028-0.025)^2 + (0.068-0.068)^2 + (0.046-0.069)^2 + \ldots\ldots + (0.011-0.011)^2} = 0.065.$$

Table 3.36 Route selection for transmission line

	Suquia river valley	Trevel mountains	Suquia valley/mountains
Distance between project and ideal $(R)^+$	0.065	0.038	0.055
Distance between each project and worst $(R)^-$	0.039	0.071	0.052
Result	0.38	0.65	0.49

The same computation yields values for the other two alternatives, 0.038 and 0.055. Applying the same procedure, $(R)^-$ values are calculated and shown in the corresponding row in Table 3.36.

Now it is possible to compute the *'closeness index'*.

$$\text{Closeness index }(\text{C.I.}) = \frac{(R)^-}{(R)^+ + (R)^-}$$

Then for the *first alternative*: Closeness index = 0.38
For the *second alternative*: Closeness index = 0.65
For the *third alternative*: Closeness index = 0.49

The best value is the highest, therefore the ranking is:

First: **Trevel Mountains**
Second: **Suquia River Valley/Mountains**
Third: **Suquia River Valley**

3.2.5.3 Case Study: Selection of Urban Alternatives Routes (Solved Using TOPSIS 'Six Sigma' Software)

To illustrate the method a TOPSIS software developed by 'Six Sigma – Statistical Design Institute (SDI) will be used.

Background Information

A City Hall's annual budget contemplates the execution of a wide portfolio of projects affecting different departments such as Public Works, Municipal Infrastructure, Public Health, Education, etc. In this example, we only analyze the situation of the Municipal Infrastructure Department, and within it, different projects such as to improve the main highway that traverses the city from North to South East, identified as Hwy. 98. It is understood that in the real case all of City Hall's projects, no matter the area they belong to, will be in the same decision matrix. This software allows for 200 alternatives and 200 criteria.

Objective:
Improve road traffic in a section of a highway that traverses the city in order to eliminate present day bottlenecks.

Alternatives:
Feasibility studies show four possible alternatives, which are:

(A) Full expansion of Hwy 98. It means a complete upgrading of the elevated highway, practically to make it anew.

(B) Widening Hwy. 98. It requires only expansion, by adding two side lanes in some sections.
(C) Build a parallel highway at street level. It involves leaving the highway as it is, and building a parallel road at street level.
(D) Build a bypass to bridge the busiest section of the highway. It means a bypass, which practically will bridge the downtown area of the city.

Criteria:
The decision matrix shown in Table 3.37, considers fifteen selection criteria. They are:

1. City Hall regional policy
   This criterion refers to City Hall policy regarding different alternatives, appraised through a 1 to 15 points evaluation, the higher the better; that is the aim is to maximize the policy, indicated in the 'Action' column. After a series of considerations where no costs are involved, subjects such as urban aesthetics are discussed, as well as the impact of alternatives on the old city, social questions, vibrations, etc. In this scale a 9 is assigned to alternative A, i.e. favourable enough. Alternative B obviously gets the lower ranking because, as a practical matter, it maintains the status quo and would not help the city. Alternative C gets a little better ranking (albeit not too high because it implies the demolition of many dwellings, which is not only costly but unpopular).
   It is evident that, from the government point of view, D is the best alternative. An expert panel estimated these values, with pros and cons for each alternative freely discussed, and then made a selection by secret voting.
2. Citizens' opinions
   For this criterion, citizens' opinion is requested through a mail survey across the city and through face-to-face interviews with people living nearby the highway. These opinions are valued from 1 to 100. The highest value corresponds to alternative C which would possibly indicate that most people prefer a better road connection.
3. Environmental and landscape subjects
   For this criterion, the alternative that gets the highest ranking from citizens is again alternative C, mostly because a new route would valuate properties along this road.
4. Economic subjects
   According to this criterion, the preferred alternative is B, since it is thought that it will generate a considerable economic activity because there will be more traffic between the city, its districts and metropolitan area. This criterion analysed the economic contribution of each alternative; it does not refer to project costs but to the potential benefits it can bring.
5. Traffic volume
   Largest traffic volume belongs to alternative D.
6. Traffic flow.
   Same.
7. Safety

Table 3.37 Decision matrix

	Alternatives ⟶	Full expansion of Hwy. 98 (A)	Widening Hwy. 98 (B)	Build a parallel Hwy. at street level (C)	Build a bypass to bridge the busiest section of the Hwy. (D)	Units	Action
	Criteria						
1	City Hall regional policy	**9**	**5**	**7**	**12**	1–15	Max
2	Citizens' opinion	**27**	**17**	**35**	**21**	1–100	Max
3	Environmental and landscape subjects	**23**	**20**	**40**	**17**	1–100	Max
4	Economic subjects	**14**	**39**	**16**	**31**	1–100	Max
5	Traffic volume	**10,000**	**6,000**	**3,500**	**12,500**	Number of vehicles	Max
6	Traffic flow	**3,000**	**2,000**	**1,200**	**4,200**	Cars/h	Max
7	Safety	**200**	**320**	**320**	**175**	Accidents/year	Min
8	Urban traffic	**6,500**	**6,900**	**6,200**	**1,200**	Number of cars	Max
9	Regional traffic	**2,500**	**1,500**	**1,850**	**13,000**	Number of cars	Max
10	Direct economic benefits	**12**	**19**	**12**	**11**	Millions of Euros	Max
11	Indirect economic benefits	**9**	**20**	**9**	**8**	Millions of Euros	Max
12	NOx content	**50**	**50**	**62**	**44**	ppb	Min
13	SO_2 content	**65**	**65**	**71**	**49**	ppb	Min
14	Building areas affected by the undertaking	**29**	**34**	**38**	**41**	ha	Min
15	Budget participation	**1**	**1**	**1**	**1**	Number	Equal

The best is alternative D, because estimates show that it will have the lowest value of traffic accidents per annum.

8. Urban traffic
   Urban traffic maximizes with alternative B, possibly because the expansion means the building of more urban access.
9. Regional traffic
   This criterion refers to benefits because there will be a direct connection between peri-urban areas of the West and North of the city. Preferred activity is the one that maximizes said traffic, which is D.
10. Direct economic benefits
    Economic benefits are apparently larger for alternative B, possibly because there will be no interruption of traffic during construction, as in alternatives A and C.
11. Indirect economic benefits
    Again, alternative B is the one that offers the greatest indirect benefits, perhaps from the influence of direct benefits.
12. NOx content
    Since this criterion aims at obtaining the minimum contamination possible, the alternative offering less contamination is D, because it will manage only a portion of traffic compared with that of Hwy 98.
13. SOx content
    Same, the best alternative is that one of lesser value, which is D.
14. Building areas affected by the undertaking
    Best alternative is the lesser that is A, because it involves least housing demolitions.
15. Budget participation
    This row is not a criterion. Unit values at the intersection of a column and a row simply indicate that all projects participate in the analysis when it is time to discuss the appropriation of funds. This is really a calculus device, but observe that its action is neither to maximize nor to minimize but expresses that the algebraic summa must be equal to a certain value and consequently the sign '=' (that is equal to a certain value) is used.

Construction of the decision matrix:
Table 3.37 shows all data.

Normalization

In order to be able to make meaningful comparisons, all the values of Table 3.37 are to be normalized, using any of different methods, for instance dividing each value in a row cell by the maximum values of that row, or by the total of row values, etc. Table 3.38 shows all scores normalized.

This is the information that is loaded in Six Sigma software. See Fig. 3.12. It has a template with options or alternatives in columns and criteria in rows, and there is also a column for criteria importance or weights. Fill the template with data from

Table 3.38 Normalized values from Table 3.37

	Alternatives →	Full expansion of Hwy. 98 (A)	Widen Hwy. 98 (B)	Build a parallel Hwy at Street level (C)	Build a bypass to bridge the busiest section of the Hwy. (D)	Units	Action
	Criteria						
1	City Hall regional policy	**0.273**	**0.152**	**0.212**	**0.364**	1–15	Max
2	Citizens opinion	**0.270**	**0.170**	**0.350**	**0.210**	1–100	Max
3	Environmental and land scape subjects	**0.230**	**0.200**	**0.400**	**0.170**	1–100	Max
4	Economic subjects	**0.140**	**0.390**	**0.160**	**0.310**	1–100	Max
5	Traffic volume	**0.313**	**0.188**	**0.109**	**0.391**	Number of vehicles	Max
6	Traffic flow	**0.288**	**0.192**	**0.115**	**0.404**	Cars/h	Max
7	Safety	**0.197**	**0.315**	**0.315**	**0.172**	Accidents/year	Min
8	Urban traffic	**0.313**	**0.332**	**0.298**	**0.058**	Number of cars	Max
9	Regional traffic	**0.133**	**0.080**	**0.098**	**0.690**	Number of cars	Max
10	Direct economic benefits	**0.222**	**0.352**	**0.222**	**0.204**	Millions of Euros	Max
11	Indirect economic benefits	**0.196**	**0.435**	**0.196**	**0.174**	Millions of Euros	Max
12	NOx content	**0.243**	**0.243**	**0.301**	**0.214**	Ppb	Min
13	SO_2 content	**0.260**	**0.260**	**0.284**	**0.196**	Ppb	Min
14	Building areas affected by the undertaking	**0.204**	**0.239**	**0.268**	**0.289**	Hectares	Min
15	Costs (× 000)	**14,231**	**8,350**	**10,896**	**16,523**	Euros	Min

TOPSIS Selection Matrix

Title: Select. of urban altern. routes
Project:
Name:
Date:

Criteria			Units	Importance	1: A - Full expansion of Hwy98	2: B - Widening Hwy98	3: C - Building a parallel Hwy	4: D - Building a bye-pass to bridge the busiest section of the Hwy	Goal	- Ideal	+ Ideal
	1	City Hall regional policy	Interval	0.08	0.273	0.152	0.212	0.364	maximize	0.152	0.364
	2	Citizen's opinion	Interval	0.08	0.27	0.17	0.35	0.21	maximize	0.17	0.35
	3	Environmental and landscape subjects	Interval	0.12	0.23	0.2	0.4	0.17	maximize	0.17	0.4
	4	Economic subjects	Interval	0.07	0.14	0.39	0.16	0.31	maximize	0.14	0.39
	5	Traffic volume	Number	0.06	0.313	0.188	0.109	0.391	maximize	0.109	0.391
	6	Traffic flow	Number	0.05	0.288	0.192	0.115	0.404	maximize	0.115	0.404
	7	Safety	Number	0.03	0.197	0.315	0.315	0.172	minimize	0.315	0.172
	8	Urban traffic	Number	0.04	0.313	0.332	0.298	0.058	maximize	0.058	0.332
	9	Regional traffic	Number	0.06	0.133	0.08	0.098	0.69	maximize	0.08	0.69
	10	Direct economic benefits	Euros	0.1	0.222	0.352	0.222	0.204	maximize	0.204	0.352
	11	Indirect economic benefits	Euros	0.11	0.196	0.435	0.196	0.174	maximize	0.174	0.435
	12	NOx Content	Ppb	0.01	0.243	0.243	0.301	0.214	minimize	0.301	0.214
	13	SOx content	Ppb	0.01	0.26	0.26	0.284	0.196	minimize	0.284	0.196
	14	Building areas affected by the undertaking	Hectares	0.06	0.204	0.239	0.268	0.289	minimize	0.289	0.204
	15	Costs	Euros	0.12	14231462	8350563	10896324	16523241	minimize	16523241	8350563
				Score	0.33	0.49	0.42	0.45			

Fig. 3.12 Capture of Six Sigma template with data and results. (Screen shot for TOPSIS selection matrix reproduced with kind permission of statistical design institute LLC)

Table 3.36, specify the action required for each criterion and immediately, at the bottom of the template, the result will be displayed.

In this case the last row shows:

Alternative (A): 0.33
Alternative (B): **0.49**
Alternative (C): 0.42
Alternative (D): 0.45

Considering the highest value, the selected alternative is (B) and the ranking is as follows:

B – D – C – A

3.2.5.4 Comments on This Model

Pros:

1. It is very easy to understand because of its simplicity and rationality.
2. It can be solved manually for small problems without too much effort; and there is effective software.
3. Other than fixing weights for criteria it is one of the methods that uses less subjectivity.
4. Projects can be added or deleted without problem.
5. It gives an idea of optimality.

Cons:

1. It does not consider thresholds.

3.3 Conclusion for This Chapter

This chapter has tried to give a glimpse of the state of the art regarding available tools that can help the decision-making process. It goes without saying that not all models have been commented on or exemplified, only those which enjoy the greatest preference by users, evidenced by the abundant bibliography on them found on the Internet. Each methodology has been exemplified with a different and actual case, albeit perhaps not with the complexity of a more complicated scenario, an aspect that is not perceived as necessary to comprehend each tool. The purpose behind these cases is to give the practitioner a clear image of what he/she needs (data) to analyze problems, illustrate the steps to set up each methodology, and exemplify which is the best solving procedure in each case, both manually and using dedicated software. Once the problem is solved – perhaps the easiest part – it starts with the analysis of the results found, together with the DM, and for this step we have commented on the necessity of performing a sensitivity analysis and the way to do it.

Finally, advantages and disadvantages have been expressed for each methodology as per the opinion of this writer, who is not endorsing any special method; all of them have good and not so good points, particular fields of application for some of their aspects, and strong and weak features. Once the practitioner knows the different models and understands their inherent difficulties, he/she will be better able to select the model that best matches his/her needs. Or if necessary, the practitioner can go deeper into the knowledge of each model, consulting specialized books and – most important – the operation manual that the vendor of a methodology offers.

As pointed out, there are many other models, as those mentioned in Sect. 9.3, where Table 9.9 lists some applications of techniques such as REGIME, SAMI, V.I.S.A, etc. This table may be useful as a guide for practitioners, since it groups 66 actual projects and constitutes a source of information, for it shows the methodology applied in different types of projects. Some projects included in this table were solved using techniques akin to Linear Programming, a system whose practical applications are discussed in some detail in Chap. 4.

References

Bana e Costa, C., Vansnick, J-C. (2001). *A fundamental criticism to Saaty's use of the eigenvalue procedure to derive priorities*. Working paper LSEOR 01.42ISBN No. 0 7530 1503 X. London: Department of Operational Research London School of Economics and Political Science.

Brans, J., & Vincke, Ph. (1985). A preference ranking organization method: The PROMETHEE method. *Management Science, 31*, 647–656.

Brans, J., Vincke, P., & Mareschal, B. (1986). How to select and how to rank projects: The PROMETHEE method. *European Journal of Operational Research, 24*, 228–238.

Brent, A. C., Rogers, D. E., Ramabitsa-Siimane, T. M., & Rohwer, H. B. (2007). Application of the analytical hierarchy process to establish health care waste management systems that minimize infection risks in developing countries. *European Journal of Operational Research, 181*(1), 403–424.

Dyer, J. (1990). Remarks on the analytical hierarchy process. *Management Science, 36*(3), 249–258.

Flament, M. (1999). *Glosario multicriterio*. Edmonton: Universidad de Alberta. Retrieved from http://www.unesco.org.uy/red-m/glosariom.htm.

Ghazinoory, S., Aliahmadi, A., Namdarzangeneh, S., & Ghodsypour, H. (2008). Using AHP and L.P. for choosing the best alternatives based the gap analysis. *Applied Mathematics and Computation, 184*(2), 316–321 (2007).

Gomory, R. (1958). Outline of an algorithm for integer solutions to linear programs. *Bulletin of the American Mathematical Society, 64*(5), 275–278.

Guerrero Padilla, N., Quijada Arévalo, M., & Guerreo Casas, F. (2000). *Aplicación del método PROMETHEE en la jerarquización de títulos de una cartera de valores*. Métodos matemáticos para la economía y empresa. Madrid: UNED.

Hinloopen, E., Nijkamp, P., & Rietveld, P. (1983). The regime method: A new multi-criteria technique. In P. Hansen (Ed.), *Essays and surveys on multiple criteria decision making*. Berlin: Springer.

Hwang, C., & Yoon, K. (1981). *Multiple attribute decision making: Methods and applications, a state- of – the – Art survey*. Berlin: Springer.

*Jaramillo, P. (2002). Métodos de análisis multiobjetivo para problemas continuos. Universidad Nacional de Colombia, Sede Medellín Módulo 4. *Journal of Multi-criteria Decision Analysis*. (Wiley). http://pisis.unalmed.edu.co/vieja/cursos/analisis_decisiones/AD_4_%20ponderantes%20y%20compromiso.pdf.

*Jaramillo, P., Álvarez, J., & Quintero, R. (2002). Ecualizador: un método integral para la decisión con múltiples objetivos. *Ingeniería del agua*, 9(2).

*Kangas, J., Puakkala, T., & Kangas, S. (2001). HERO heuristic optimization of multi-criteria forestry decision analysis. In *The analytical hierarchy process in natural resources and environmental decision making*. Dordrecht: Kluwer.

Lootsma, F. (1999). The French and the American school in multi-criteria decision analysis. In F. Lootsma (Ed.), *MCDA in the hands of its masters*. Dordrecht: Kluwer.

Munda, G. (2004). Métodos y procesos multicriterio para la evaluación social de las políticas públicas. *Revista Iberoamericana de Economía Ecológica, 1*, 31–45.

Pomerol, J.-C., & Barba-Romero, S. (2000). *Multicriterion decision in management: Principles and practice*. Leiden: Springer.

Roy, B. (1968). Classement et choix en présence de points de vue multiples (la méthode ELECTRE). *Revue Francaises Informatique Recherche Operationnelle, 2*(8), 57–75.

Rogers, M., Bruen, N., & Maystre, L. (1999). *ELECTRE and decisions support. Methods and applications in engineering and infrastructure investment*. London: Kluwer.

Saaty, T. (1980). *Multicriteria decision making – The analytic hierarchy process*. New York: McGraw-Hill.

Van Delft, A., & Nijkamp, P. (1977). *Multi-criteria analysis and regional decision-making*. Leiden: H.E. Stenfert Kroese, B.V.

Vincke, P. (1992). *Multicriteria decision aid*. New York: Wiley.

Wallenius, J., Dyer, J., Fishburn, C., Steur, R., Zionts, S., & Deb, K. (2008). Multiple criteria decision making, multiattribute utility theory: Recent accomplishments and what lies ahead. *Management Science, 54*(7), 1336–1349.

Zeleny, M. (1982). *Multiple criteria decision making*. McGraw-Hill: New York.

*Indicates suggested reading material not mentioned in the text.

Chapter 4
Linear Programming for a Single Objective

Abstract This chapter is dedicated in its entirety to Linear Programming, a well-known mathematical procedure which has an enormous diffusion in hundreds of applications around the world. This technique, using a practical example, is explained in a way for everybody to understand it. It aims at making the DM aware of how to use this tool, and more important, how to interpret its results. Linear Programming as is explained here deals with a sole objective which is common in many applications and in different fields. Its greatest advantage can be synthesized on three counts: (a) It permits one to approximately represent an actual situation – no matter its nature – in a mathematical context, that allows for applying an algorithm to solve it, (b) it yields a unique and optimal solution, and (c) it lets to perform an extensive analysis of "*What if….?*" scenarios which is a valuable tool for sensitivity analysis.

Keywords Linear programming • Graphic solution • Simplex method • Multicriteria • Objective function

4.1 Brief Theory on Linear Programming

Linear Programming (LP) works in the domain of linear algebra, especially in matrix algebra. The concept of the method is rather simple. It starts with the data or decision table, which contains, as stated earlier, the *set of scores of each alternative* corresponding to each criterion, identified as $\mathbf{a}_{ij}$ and conforming matrix **'A'.** The *variables or alternatives* are represented as $\mathbf{x}_j$ (expressed as a vector **X**), and the *thresholds* ($\mathbf{b}_i$), make out vector **B.** We have already worked with these elements individually, and now is the time to link them in a matrix format, as follows:

$$\mathbf{A \cdot X = B}.$$

N. Munier, *A Strategy for Using Multicriteria Analysis in Decision-Making: A Guide for Simple and Complex Environmental Projects*, DOI 10.1007/978-94-007-1512-7_4,

Our unknowns are the **X**, i.e. the solution vector of proposed projects or alternatives. What we wish to do is to determine which of those proposed alternatives is/are the best for the job. The answer lies in solving this matrix equation for **X,** then:

$$\mathbf{X} = \mathbf{B} / \mathbf{A}.$$

And with a linear objective function[1]

$$\mathbf{Z} = \mathbf{C} \cdot \mathbf{X}$$

4.1.1 Modeling Actual Cases

It is seldom that in actual cases diverse alternatives of a project are not related with each other (for instance, diverse options for highways, transmission lines, industrial locations, investments, etc.), and from that point of view, complementarity or substitution between alternatives was investigated a long time ago by Weingartner (1966). Therefore, our mathematical model has to represent as faithfully as possible these conditions, and also be able to express the result in different ways, depending on the nature and characteristics of the problem, for instance:

- In fractional values for each alternative, in this case the larger the better; and this is usually the case when the result may be also expressed as a combination or blend of projects, all competing for the same resources. For instance, a battery of urban projects including sewage, health, environment and social projects, competing for funds, manpower, land, etc.
- In integer values, that is no fractions, because obviously it is not possible to build 0.35 of a hospital, 2.84 houses, 6,787.9 cars, etc. They must have integer values.
- In binary values. Very common, when a '1' means 'Yes' and a '0' means 'No'. This is a very usual scenario when there are exclusive projects under consideration, that is, it is one or the other, but not both.

It is simple to state these conditions in a LP model. For instance a fractional result conditions may be expressed as:

$$\sum_{i=1}^{k} xj \leq 1.$$

This is the case when we want a ranking of alternatives as a solution.

If a project X_B depends on the completion of project X_A, the following relationship can be established:

$$X_A \leq X_B.$$

[1] There could be a non-linear objective function, originating non-linear programming problems. Pioneered and investigated by Kuhn and Tucker (1950).

This is for instance the case when there are two projects involving the same product (say clothes washers), and one of them consists in performing a market study to measure the demand for that appliance (X_A), and the other for the construction of a plant to manufacture it (X_B). That is, it is possible to inform the model that Project X_B (planning plant construction) cannot start if Project X_A (market study), has not been approved.

It can also happen that there are two projects X_C and X_D which depend on the completion of X_A and X_B, and in this case this is indicated by:

$$X_A + X_B \leq X_C + X_D.$$

There also could be the case that there is a nonlinear restriction and in that case:

$$X_A . X_B = 0.$$

Sometimes, for several reasons, it is required that a particular project must be part of the solution, for instance a project already under execution, i.e. the solution is forced to accept something already established. This case is expressed as:

$$Xj \geq 0.$$

Occasionally there are very strict constraints, such as in the case that funds have to be spent totally in a fiscal year (otherwise they have to be returned); in that case:

$$\sum_{j=1}^{n} aij \cdot xj = 1.$$

The above formula is also used when a ranking or ordering must be expressed as a percentage or when there is the need to indicate which projects to consider.

As we see, the methodology allows for large latitude in representing real life conditions.

4.2 Solving a Multicriteria Problem Through the Graphic Method

To clarify matters let us start with an example in which there are only two projects.

4.2.1 Case Study: Selection Between Two Sources of Renewable Energy (Solved Graphically)

An industrial developer wants to take advantage of tax benefits for capital investment in renewable energy projects. From that point of view, he is contemplating two alternatives:

(a) Using solar energy panels or dishes (SE) and
(b) Employing photovoltaic cells (PV).

In the first project, energy is produced when sunrays impinge on a battery of slightly parabolic mirrors tracking the sun from sunrise to sunset. The sunrays coming in parallel waves strike a mirror, which because of its geometry concentrates them into a focal point located in a central tower (see picture) (the same phenomenon produces a magnifier lens when receiving the sun's incoming rays and can burn a piece of paper located at a certain distance).[2] In this case, the concentrated heat in the tower reheats a salt solution at a very high temperature. The thermal energy is then transmitted to water making it boil, and generating high-pressure steam used to move a steam turbine, which drives an electric generator. To illustrate the system, Fig. 4.1 shows an example of this scheme (mirrors and tower), albeit this actual installation is not related whatsoever neither with this case nor with the data.

Fig. 4.1 Mirrors and tower in San Lucar, Seville, Spain

[2]Legend has it that Archimedes set Roman warships afire using '*a burning glass*' during the siege of Syracuse.

The second scheme uses panels of photovoltaic cells, which convert sun radiation directly into electrical energy. Both alternatives generate clean and inexhaustible energy and have different characteristics regarding efficiency, cost per kW produced, land use, etc. The entrepreneur has purchased a large track of land in sunny country where either of the two schemes can operate isolated or in tandem, i.e., in one arrangement the system can function only with SE, or only with PV, or with a combination of both. In this very simple example, the DM and staff considered four criteria indexes, to which both systems must comply. Each alternative possesses a value or score for each index obtained from technical studies.

The problem is now to settle on the best solution when considering that the objective calls for cost minimization. The first step is to build the decision matrix as shown in Table 4.1.

Table 4.1 Decision matrix for the renewable energy case

Cost index:	0.72	0.68	(Minimize)	
Alternatives or variables →	Solar energy (SE)	Photovoltaic (PV)	Action	Threshold
Criteria				
(A) Efficiency index	**0.85**	**0.75**	Maximize (≤)	**1**
(B) Financial index	**0.78**	**0.98**	Minimize (≥)	**0.84**
(C) Land use index	**0.92**	**0.65**	Maximize (≤)	**0.94**
(D) Generated energy index	**0.99**	**0.60**	Minimize (≥)	**0.80**

Observe that scores are quantitative and real values too are derived from technical data. This scenario can be put into a mathematical context (this step is not necessary in practice but it is done here only for the purpose of clarifying concepts). Equations are as follows:

(A) Efficiency index: $0.85\ \text{SE} + 0.75\ \text{PV} \leq 1.00$
(B) Financial index: $0.78\ \text{SE} + 0.98\ \text{PV} \geq 0.84$
(C) Land use index: $0.92\ \text{SE} + 0.65\ \text{PV} \leq 0.94$
(D) Total generation index: $0.99\ \text{SE} + 0.60\ \text{PV} \geq 0.80$

That is, in the first equation it means:

Solar energy alternative contributes with a score of 0.85 to the Efficiency index, while photovoltaic contributes with 0.75. When the algorithm finds the values for SE and PV, then the first equation must be satisfied. That is, the summation of the products of scores by these found values must be lower than 1 (the threshold), and hence the sign '<' that is '*lower than*', for if the criterion refers to efficiency it is obvious that the combined efficiency must be below 1, because efficiency cannot have a value larger than 1.

Surely the reader will wonder where these values come from. Let us do a little analysis for the scores of both the objective function and the criteria, not as a technical digression but to illustrate which considerations can and must be taken into account for defining scores, wherever they belong.

4.2.1.1 Objective Function Scores

The objective function calls for a minimization of costs and its values suggest that generation with SE (0.72) is more expensive than PV (0.68). These values are the result of considering diverse issues, for instance, in the SE they could be:

(a) Acquisition of mirrors and tracking devices,
(b) Tower construction,
(c) Acquisition of generation equipment,
(d) It needs to use part of the generated energy to operate the tracking devices,
(e) This system has the advantage of generating alternating electrical current, which directly injects into the electric grid.

Considering now the PV system, we have:

(a) PV panels are more expensive than mirrors,
(b) Probably efficiency is reduced in a larger amount than in SE because of dust,
(c) The system generates direct current; therefore it needs conversion into alternating current through devices called '*power converters*'.

As can be appreciated, this is a complex issue, but its evaluation is essential to define the scores for the objective function. However, these values could be far from certain and then the result will reflect this lack of exact appraisals. Because of that, as in other problems, once they are solved it is recommended to perform a sensitivity analysis to appraise how significant changes in these scores affect (or do not) the result, which indicates the solution's degree of stability.

4.2.1.2 Criteria Characteristics

Look at the first criterion (*Efficiency index*) to be maximized, which means getting the maximum efficiency, which has a top value of 1. Notice that other than the fact that both systems work under completely different physical principles, in general solar energy has a higher efficiency because it can be used from sunrise to sunset and after, because heat can be stored during the day in special devices and then utilized at dusk and part of the night. It is not the same with PV because energy generation stops at sunset.

The *financial index* calls for minimization, that is the sum of the products must have, as a minimum, a value no less than 0.84, which could represent some sort of return on investment (obviously, not as a percentage).

Similar considerations are valid for the third and the fourth criteria. In the *Land use index* the inequality says that whatever the blend of systems, the land use should not exceed 0.94, which could represent for instance the maximum size of the plot of land the developer has.

The fourth criterion, the *generated energy index*, relates to total energy generated, and similarly, the inequality indicates that, whatever the blend of technologies used, the energy should be as a minimum 0.80, which could represent the economy of scale for this class of undertakings. There could also be a criterion indicating that whatever the blend of technologies used the cost per KW must not exceed a certain value, which is normally established by the market; otherwise the energy produced could be not competitive.

This only purpose of this brief commentary has been to illustrate how a mathematical model can approximately mimic real life situations. In this example, we have examined:

- The cost scores for the objective function as well as the threshold,
- Scores justification for the efficiency index, as well as the threshold,
- Scores for financial, land use, and total generation indexes, and corresponding thresholds.

As can be seen, there is no room for subjectivity here, and this is one of the strongest points of LP; however, because actual values are used, it is possible that the reader will wonder what is then, in this case, the DM function? Well, primarily it is he/she who decides what criteria to use, since they surely assume that the alternatives are technically sound. On the other hand the DM, even if he/she cannot possibly argue on technical subjects, must usually deal with other matters which do not have the benefit of having technical or true scores, as mentioned above.

For instance, most probably, the project will involve issues that affect people; this data normally comes from surveys, polls and consultations. This information usually needs a certain level of interpretation from the DM as well as its linking with the project environment boundaries, and it could be subject to DM preferences. Besides, the DM may or may not be in agreement with the thresholds and will ask for changes. For instance, in a project that calls for a plan to protect a sensitive heritage, the economic criterion might imply establishing fees for visitors. Here the DM can have a strong voice, based on his/her knowledge, experience and interests. In conclusion, remember that it is the DM who accepts or rejects a solution given by any model and might choose instead a second or even a third alternative in the ranking.

4.2.1.3 Criteria Actions

In this simple example there are only four criteria, with maximization and minimization actions, and of course they can be combined, but in an actual case there could be hundreds, for instance related to aspects such as:

- Available funds (MAX),
- Damage to the environment in the production of mirrors and PV cells (MIN),
- Decrease of efficiency due to dust accumulated in mirrors and plates that reduce sun ray action (MIN),
- Degree of satisfaction or dissatisfaction of neighbourhood people with this project (MAX or MIN),
- Economic benefits brought by the project (MAX),

- Effect that the accumulated heat might have in the environment (MIN),
- Opportunity cost because of the current use of land for industrial purposes but which is also suitable for cattle raising (MIN or MAX)
- Probability of damage because of hail (MIN),
- Reliability of the system due to cloudy days, etc.(MAX).

Of course there could also be many more alternatives, such as for example installing wind turbines or combining heat with a nuclear energy station, or whatever other arrangement.

Let us come back to solving our problem now that the decision table and the objective function have been explained. How do we solve this problem? *That is, how do we find the best alternative or combination of alternatives that* ***minimize the total cost***? This is the objective and the question that the DM formulates. Since there are only two variables or alternatives, the problem is solvable graphically. To do that, consider a coordinate system formed by axis SE and PV and represent by lines A, B, C and D the system of inequations. See Fig. 4.2.

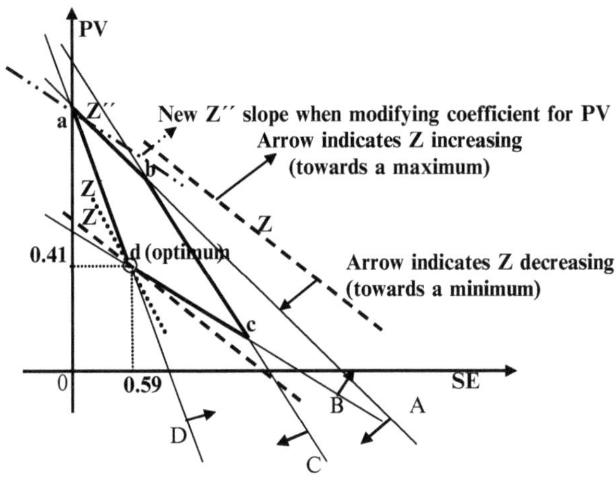

Fig. 4.2 Graphic representation of the decision matrix for the renewable energy case

As will be noted, criteria represented by inequalities A, B, C and D are denoted by straight lines and have a field of validity according to their signs. Thus, inequality A is valid in an area below this line and limited by axes SE and PV. When these conditions are considered for all criteria, a polygon a-b-c-d-a is formed. Within it are all the *feasible solutions* of the problem, but the efficient ones are only at the vertices a-b-c-d.

4.2.2 Graphic Solution

The objective function **Z** (dashed line) – which also is a linear equation – can be displaced parallel to itself until it becomes tangent to one of the vertices of the polygon.

There are two vertices where it can become tangent, **'a'** and '**d**', but 'a' corresponds to a maximum, therefore since the objective calls for a minimization, the solution is at vertex 'd'. This is the best and optimum solution. As per this result, the best solution is a combination of alternatives that is a 59% of total output must be generated by SE and 41% by PV. Notice that the model chose a larger value for alternative SE even though it is not the cheapest, and this is because other factors are also playing in the selection, not only the cost.

4.2.3 Changes in the Objective Function

Notice that if for some reason it is necessary to adjust or change the coefficients of the objective function Z, this line will change its inclination for instance as in Z′ (dotted line). However, observe that for a while Z can pivot on 'd', meaning that the solution holds even for relatively important variations in Z coefficients, and this is important because it ensures that there is stability in the solution found. The variation can be in only one coefficient or in both at the same time, and with one increasing and the other decreasing, or both varying in the same sense. Of course, this is not a norm, and in many situations a slight variation in one or two of the Z coefficients can change the solution, but knowing about this circumstance is very important for the DM, especially in projects where there is a good deal of uncertainty.

Nevertheless, if there are large changes in one or in both cost coefficients, for instance, if the coefficient for PV drops from 0.68 to 0.43, Z will change its inclination and will become a Z′′ (dash and dots line), tangent to vertex **'a'** and then offer a new solution, as shown in Fig. 4.2, consisting in developing only the PV scheme. This analysis is very significant because it allows the DM to examine what happens if the cost index for PV drops due to technological advances or mass production economies, or conversely, if the index increases for SE. Note, besides, that the original solution found is stable since there is a need for a sharp drop of more that 36% in the cost of PV for the solution to change.

4.2.4 Criteria Importance

Observe that criteria B and D are the only ones which define vertex'd'. Consequently, these two criteria are the most important of the four considered. Why are they important? Because their variation can modify the solution reached.

Effectively it can be verified that changing, for instance, the inclination of D or its position, will change the coordinates of 'd' and will alter the values found before. How can the inclination change? It can change if we modify the scores of the alternatives (even with only one of them) corresponding to criterion D, for instance switching the score of PV from 0.60 to 0.55, as if for instance it has been found that the former value is too high.

4.2.5 Change in the Thresholds

How can the position change, even holding the inclination? It is done by changing the threshold. Assume for instance, that the developer thinks that the threshold for criterion D must be higher, say 0.90 instead of 0.80 (Table 4.1). The line for criterion D displaces parallel to itself toward the right and in so doing changes the coordinates of vertices and thus the solution. This is another very important feature and in fact the effect on objective Z of one unit change in any criterion creates a '*shadow price*' or '*marginal value*' for the criterion, and quantitatively measures its importance.

4.2.6 Conclusion of This Example

This example illustrates the method and points out the quality of the information the system can provide to the DM. Now what happens in a situation where we have say seven variables or alternatives and 25 criteria? Naturally, it is not possible to represent this problem in a plane because it has seven dimensions. However, the model would resolve it in exactly the same way as explained but using mathematical dimensional spaces instead of the maximum three dimensions that we posses in our world.

To proceed in this more realistic scenario, the Simplex[3] method is used but in its analytical format which is then resolved using dedicated software such as the 'Solver' which is a standard add-in of Excel. This very powerful software can resolve in seconds very complex problems with hundreds of alternatives and criteria, provided, obviously, that the problem is correctly set up. If there are mathematical errors the Solver will signal that, but of course, the software cannot tell us if the problem was or was not well built or structured.

It also could be that the Solver indicates the lack of a solution to the problem posed, and that usually happens when there are contradictory criteria or if effectively the number of restrictions is so large that there is no solution.

The Simplex is an algorithm developed to resolve Linear Programming problems, regardless if they are small, large or very large, simple or complex. Based in matrix algebra, it is in reality a simple procedure and even amenable to solution by hand, albeit this choice is not advisable.

To finish this comment, it is necessary to mention that we have been dealing with a very simple problem, because it has only one objective function, and thus, as it has been explained, the model yields an optimum solution. However, in most real life situations there are normally more than one objective. For instance, the DM may want to consider four different objectives for a problem, such as:

(a) To get the maximum benefit,
(b) To incur the minimum investment,

[3] Algorithm developed by George Dantzig to solve Linear Programming problems.

(c) To minimize damage to the environment,
(d) To maximize resources use.

This is called a '*multiobjective and multicriteria decision problem*'.

Linear Programming cannot resolve this type of problem. To be sure, it is not for lack of effort, since there are some models using LP that find a solution by applying the Simplex method, e.g. '*GOAL PROGRAMMING*', proposed in the 1950s (Charnes and Cooper 1961), (See also Steuer 1986; and Lee 1972), and most recently '*Compromise Programming*' (Ballestero 2007), and others which in general do not find an optimal solution – an impossible task most of the time except for small problems – but which furnish a set of solutions that are considered satisfactory and from which the DM can choose.

This inability of LP to solve problems with many objectives was the reason to develop SIMUS, explained in Chap. 6. However, SIMUS does not necessarily find an optimal solution; based in the Simplex method delivers a set of satisfactory solutions with a degree of uncertainty much lower than other models and working with any quantity of alternatives and criteria.

4.2.7 Integer Solutions

The example posed illustrated a case that allows a blend of alternatives, and it does not call for a discrete answer. However, most problems aim at getting a solution specifying that only a project or alternative must be pursued, not a mix of them, and it usually involves a dichotomy or a mutually exclusive response regarding alternatives or projects, such as, *build* or *not build*, *location A* or *location B*, *alternative D* or *alternative F*, etc. Assume as an example that a large multinational corporation in the informatics business decides to install a software development centre and considers, taking into account a set of features, that there is a potential location in three cities of three different countries. However, it is not convenient to have a number of people working in a country and some in another, so it is desirable to have a unique location.

What is the objective here? To have the minimum costs, or the maximum benefit or perhaps to maximize the specialized work force needed for this venture. This problem is solved in LP with the same ease as the example above proposed; the only condition is to specify in the Solver that the solution must be of the binary type, i.e. '1' or '0'. That is, if location B is selected when competing with locations A and C, and if Solver is required to deliver a binary sort of solution, then it will show a '1' for location B, and '0s' for locations A and C.

4.2.8 Comments on This Model

Pros.

1. LP is the only methodology that can supply optimal solutions in problems with only one objective function and for small problems up to three objectives.

2. LP allows a large array of analysis based on the ease of performing a sensitivity analysis. As a matter of fact, when solving a LP model, the solution of its dual appears, which supplies all the shadow prices (See Sect. 5.3).
3. LP does not need any weights for criteria (although it allows use of them if necessary), which is the largest source of subjectivity in models such as ELECTRE, PROMETHEE, AHP and others.
4. LP allows mathematical representation of some real situations, as for example:

- Precedence
  Project A must precede project D. Assume for instance that project D is the development of a new car, while project A corresponds to designing the new model; obviously D must be done just after A is finished, and not only that, both must be executed,
- Simultaneity
  LP can work with several projects simultaneously. For instance a City Hall can have a portfolio of 250 projects of different nature, and for diverse sectors such as:

  – Social welfare: Plan to diminish street accidents which in turn presents several options,
  – Infrastructure: Several projects, like repairing 2,000 m of a sewage trunk; building a new water treatment plant; paving 478 streets, etc.,
  – Education: Maintaining 34 school buildings and constructing two buildings for high schools,
  – Cultural: Building a new public space in the downtown area,
  – Environment: Several projects such as increasing green space up to 12 m^2 per habitant; incrementally increasing to 60% of paper recycled; reducing by 10% SOx contamination, etc.

This can be done because all these projects have several things in common, mainly resources. These resources can be for instance: the municipal budget, City Hall workers and staff, demands from society, road equipment, medical personnel, number of teachers, etc. If all these projects and subprojects are placed together in columns, forming a large decision table, all criteria and restrictions in rows, and the respective projects or alternatives scored for each criterion, the problem can be solved – if, of course it is solvable.

What could be the objective in this case? If there is only one objective function, this could be maximization in the use of budget money, or maximization in quality of life, or minimization of damage to the environment, or maximizing quality of services, etc.

Another example could be to select several options related to the construction of a series of dams in several parts of a country. The objective could be the maximization of the electrical output, or a minimization of electrical blackouts, etc., and subject to a large series of issues, as for instance:

– Areas to be flooded
– Available funding for the whole project

- Avalanche possibilities
- Conservation of archaeological ruins
- Construction camps
- Difficulties of working in a particular site
- Economic viability
- Electricity transmission distances
- Existing project under construction in the same area
- Geological faults
- Logistics problems
- People's opinion on each undertaking
- Project/alternative technical difficulties
- Protected fish population
- Quantity of people to be eventually relocated
- Risks
- Water level in one lake affecting output in other hydroelectric plant located upriver

5. Freedom for the DM
   The DM can place his/her own values in the threshold, i.e. the model permits establishing the limits that he/she considers appropriate.
6. Software availability
   There are a large number of commercial software packages to use, all of them working under the same principle, to solve this kind of problem.

Cons.

1. The main drawback of Linear Programming is its lack of capacity to solve multiobjective problems, as commented above.
2. LP assumes linear relationships which is not always true in the real world.
3. LP is not easy to understand because it involves working with concepts of matrix algebra, a subject with which most DMs are not familiar.
4. Some researchers say that thresholds in LP limit the DM's ideas and free-thinking. In reality, it is not so, for what the thresholds do is to make the DM aware of existing restrictions. There is nothing limitless in our world, and resources, technical factors, capacities, etc. are not the exception, consequently, criteria must have natural or imposed limits, to take into account these boundaries.
5. LP works with cardinal values and thresholds. As a result certain criteria where subjective expressions are used such as 'poor', 'good', 'efficient', 'dangerous', 'better', etc, can not be utilized in LP, the same as in other programs. However, SIMUS (Chap. 6), allows their use after converting these subjective expressions into cardinal values using a simple scale from 1 to 10, the larger the better (the same as in other programs). Since normally these subjective criteria do not have thresholds, the DM can establish them, or else, use SIMUS property feature, which allows for these thresholds determination, if the cardinal system is normalized.

4.3 Brief and Concise Comment on the Procedure for Solving a Multicriteria Problem Through the Simplex Method

As was exemplified, criteria become linear inequalities, treated as that, while the number of variables in a problem corresponds to the number of alternatives. Consequently, a two alternatives problem has two variables or two dimensions, and can be represented and solved in a plane, as explained. When the problem has many alternatives, perhaps in the hundreds, there are correspondingly hundreds of mathematical variables or dimensions and the problem naturally cannot be represented in a plane, but it can be solved analytically.

Two alternatives constitute a coordinate system in a *plane* (a two- dimensional space); three alternatives constitute a coordinate system in *space* (a three-dimensional space); more than three alternatives constitute a coordinate system in a *multi-dimensional space*, which we cannot even imagine. Criteria are defined by the alternative scores whatever their dimension. For instance, in two alternatives problems, with say five criteria, the lines representing these criteria can be drawn in the two-dimensional space. In so doing they form a polygon, as shown in Fig. 4.2, which is called a '*solutions polygon*', because all the feasible solutions of the problem posed are contained within that polygon.

In the case of three alternatives or three variables, there is a three-dimensional space, where each coordinate axis corresponds to each variable. In these circumstances, the criteria are no longer lines in a plane but planes in a three-dimensional space. In that case, planes combine forming a polyhedron, and all feasible solutions will be within that polyhedron, also called a '*polytope*'.

When there are say, 17 alternatives, then the problem will have 17 variables and there will be a coordinate system forming a space of 17 dimensions, impossible for us to imagine but very possible to describe in mathematics. The criteria will then be hyper-planes of 17 dimensions and their intersections will form a hyper-polyhedron or polytope, where all feasible solutions are contained. A solution may then have coefficients for each one of the 17 alternatives, although this is not mandatory, that is, a solution might have only coefficients for say 10 alternatives and zero coefficients for the balance.

We have said that all feasible solutions are within the polygon, the polyhedral or the hyper polyhedron, or 'polytope' in general. What does the word '*feasible*' mean? A feasible solution is that result that complies with all the criteria of the problem. An interesting and very important property is that in the border of this polytope, which can have many vertices, are located all the *efficient solutions or non-inferior solutions* of the problem.

Now a problem arises, if we have so many efficient solutions how do we know which are the *best of them*? The objective function allows finding it. What is the objective function? It is a linear equation with as many variables as alternatives. Each variable has a coefficient or score; consequently, the result shown by the objective function will be the summation of the product of each score by the amount found for the variables when solving the problem, and identified as Z. In the example of 17 variables, the objective function will be the sum of the product of each score for variable X_j with the amount or value found as a solution of the problem for said variable X_j.

It is worth noticing that the objective function is perhaps the least significant consideration, since the most important feature is the compliance of criteria, while the purpose of the objective function is simply to pinpoint the best possible solution (Dodgson et al. 2002). This same publication adds, "Sometimes it is not obvious if an aspect of a particular problem must be represented as a constant or it should be relegated to an objective function."

In addition:

"In actual problems it is common enough to have several objectives at the same time without an obvious way to decide which the objective function should be and the balance represented as restrictions". It is also interesting to cite a comment from Malczewski (1999), when says "The objectives are functionally related to/or derived from a set".

Therefore, once the objective function is defined, the vertex where it tangents the hyper polyhedron is the best solution of our problem, and consequently cannot be improved, that is, is the optimum, and this is valid either if the scenario calls for maximizing benefits or minimizing costs. The tangency between the objective function and the polytope depends of course of the inclination of the objective function hyper plane, defined by its coefficients.

Consequently, usually the objective function '*swings*' on a vertex and it is then possible to vary its coefficients without varying the solution reached (See Fig. 4.2 in Sect. 4.3). Observe that Z (in dashed line) can swing from being parallel to B to being parallel to D around vertex 'd'. This is a very useful property for performing sensitive analysis, which is practically mandatory in decision problems. However, most possibly, changing the coefficients for whatever reasons, will change the inclination of the objective function large enough to make tangency with another vertex, and giving then another solution, as also shown and explained in that Fig. 4.2.

To better explain this methodology the same example of the two renewable energy projects that was utilized to explain the graphic method will be used.

Steps:

First:
Build the decision matrix as shown in Table 4.1.

Second:
Establish the inequalities A, B, C and D as shown in Sect. 4.3.

Third:
The Simplex algorithm eliminates inequalities by adding slack variables, which of course are not real, but used as the starting point.

Fourth:
If we represent these variables in these artificial axes, of course the first solution in this example will be at the intersection of axis SE and PV, which is 0. Thus $Z=0$

Fifth:
In this step the Simplex looks for the two real variables (SE and PV), and selects, through an algebraic rule, which of them will improve the solution best, and chooses alternative PV generating vertex 'a' (Fig. 4.2). That is, selects, out of the two, which is the variable that will enter into the solution to replace one of the slack variables.

Sixth:
Now the Simplex determine through an algebraic rule which is the slack variable which must abandon the system, in order to maintain the two dimensional scenario.

Seventh:
The Simplex repeats steps fifth and sixth, computes Z and finds out if this is the best solution; if it is not goes to the next step.

Eight:
The Simplex eliminates the other slack variable and introduces variable SE determining vertex 'd' and stops here because there is no more room for improvement, and that is the final solution.

In an actual problem, this sequence repeats thousands of times in a process called 'iteration' until reaching the optimum. This way the model can work with hundreds of variables and hundreds of criteria, but with only one objective function. This was the outline of the model, now a case will show how it works.

4.3.1 Type of Information to Input into the Model

Regarding the expected solution, it is necessary to inform the model about:

- Nature and type of project. Assume for instance plans to manufacture diverse car models, and being the purpose of the analysis to determine how many units of each model. The outcome must obviously be in integers, since it does not make sense to find a result for manufacturing say 259.67 model 'A' cars, and 410.88 model 'B' cars.
- If the problem demands the result to express the execution or not of a project, alternative or program, then the solution must be in binary values, that is, it must indicate respectively, through '1s' and '0s' which alternatives have been selected and which not.
- Very often the DM needs projects or alternatives ranked per importance, i.e. a solution in percentages must be requested, the higher the better.

Remember that we are using the example for two renewable energy project (Sect. 4.3), and then Table 4.2 replicates Table 4.1.

LP uses Table 4.2 as the base for calculations. As said, in this type of problems it is necessary to identify the alternatives that best comply with the improvement of the objective function (either maximizing or minimizing), and matching all criteria

Table 4.2 Decision matrix for the renewable energy case

Cost index:	0.72	0.68	(Minimize)	
Alternatives or variables →	Solar energy (SE)	Photovoltaic (PV)	Action	Threshold
Criteria				
(A) Efficiency index	**0.85**	**0.75**	Maximize (<)	1
(B) Financial index	**0.78**	**0.98**	Minimize (>)	0.84
(C) Land use index	**0.92**	**0.65**	Maximize (<)	0.94
(D) Generated energy index	**0.99**	**0.60**	Minimize (>)	0.80

or restrictions simultaneously. Consequently, alternatives in columns,[4] are unknowns and as such identified with 'x'. Thus, alternative A will be variable x_1, and alternative B will be x_2. Naturally, we can also continuing using A and B, but it is preferable to utilize the mathematical notation, which identifies the x_j as unknowns of the problem. The subscript 'j' indicates the successive values that the alternatives can take along a row, and in this cases $j = 1,2$.

Note that Table 4.2 admits any blend of units in the various rows, but obviously provided that they are homogeneous in a particular row. To show the correspondence between the algebraic system and the decision table observe that, for instance

For criterion (A):

$$0.85x_1 + 0.75x_2 \leq \text{To a certain quantity which is known as RHS (Right Hand Side), or simply } '\textit{requirements column}'. \text{ As the reader can guess, this is the threshold. In this case, 1.}$$

For criterion (B):

$$0.78x_1 + 0.98x_2 \geq \text{ To a certain quantity...... In this case, 0.84.}$$

For criterion (C):

$$0.92x_1 + 0.65x_2 \leq \text{ To a certain quantity...... In this case, 0.94.}$$

For criterion (D):

$$0.99x_1 + 0.60x_2 \geq \text{ To a certain quantity...... In this case, 0.80.}$$

Notice that in criteria where actions are expressed as maximization as in row (A), the corresponding inequation sign '≤' (less or equal than a maximum value) is used.

[4] In most multicriteria publications the inverse system is normally used, in which the alternatives are in rows while the criteria are in columns, but naturally, this difference does not alter results. It is probable that in LP, alternatives are in columns due to the set of 'x_j' variables representing them, which are always in columns in mathematics literature.

In others like in criterion (B), the purpose is opposite, i.e. to minimize and in that case, the sign '≥' (greater or equal than a minimum value) is utilized. There could be a criterion that calls for equality, and which is then an equation – not an inequality – which says that the algebraic sum of all values must be equal to something, and for that reason the sign '=' is used (not seen in this example).

This equality sign is often used in this methodology, besides the purpose expressed above, for instance when it is necessary to indicate that the total of funds utilized must be equal to a certain value, no less, because then there will be unused or idle funds, but not larger either because it would mean lack of funds.

4.3.2 Using the Software to Solve the Problem

Once thresholds are established for each criterion the system is ready for solving using the software 'Solver'. This Solver button appears in the same Excel spreadsheet (if it does not appear is because it was not invoked).[5] If we solve this problem using the Solver the result will be SE = 0.59 and PV = 0.41, which is the same result reached in the graphic example portrayed in Fig. 4.2. It is interesting that the Solver shows in this case, the dual values for criterion B and C, the two constituent criteria that conform the solution, as follows:

$$B = 0.48 \quad \text{and} \quad D = 0.35$$

These are *marginal values* and indicate how much the objective function Z and the solution vary when there is a unit change in each criterion. From this point of view it is evident that criterion B is more important than criterion D since the former produces a larger variation that the latter. They are the equivalent of weights for criteria used in conventional models. However, observe that these values are not subjective as weights, because they are a consequence of the values originally imputed to the model. In this way, subjectivity attached to criteria weights are avoided here.

4.4 Conclusion of This Chapter

This Chapter is merely an introduction of the main subject of this book, which is the use of LP as a technique able to deal with decision problems with multiple objectives. Its intention was to provide some theoretical background and to indicate the way the decision matrix is prepared. There is no doubt that it provides just an outline of this methodology, based in matrix algebra, discipline with which many DM are not familiar, and that was the reason in developing a graphical example, which albeit simple served

[5] To invoke, go to the large button located upper left, then hit Excel options, next Excel complements, after that, mark in the box for Solver and finally, back in the Excel spreadsheet hit 'Data'. A question mark along with the word 'Solver' will appear in the upper right corner with the word 'Solver'. Double click it and you are on.

to illustrate the method and especially to understand it. After this introduction, a proposed actual problem illustrated the model; in so doing, this Chapter offered a reader a glimpse of the mechanics for solving decision problems through LP.

References

* Avriel, M., & Golany, B. (1996). *Mathematical programming for industrial engineers*. New York: Marcel Dekker.

Ballestero, E. (2007). Compromise programming: A utility-based linear-quadratic composite metric from the trade-off between achievement and balanced (non-corner) solutions. *European Journal of Operational Research, 182*(3), 1369–1382.

* Calderón, S., & González, A. (1995). *Programación matemática*. Spain: Universidad de Málaga Departamento de Economía Aplicada (Matemáticas). Retrieved January 9, 2010, from http://www.uma.es/

Charnes, A., & Cooper, W. (1961). *Management models and industrial applications of linear programming*. New York: Wiley.

* Conde Sánchez, E., Fernández García, F., & Puerto Albandoz, J. (1994). Análisis interactivo de las soluciones del problema lineal múltiple ordenado. *Question, 18*(3), 397–415.

* Diwekar, U. (2007). *Introduction to applied optimization* (2nd ed.). Dordrecht: Springer.

Dodgson, J., Spackman, M., & Pearman, A. (2002). DTLR (Department of Transportation, Local Government and Regions). *Multicriteria analysis manual*. London: National Economic Research Associates and DTLR members.

* Elliott, D. (1978). *Planning investments in water resources by mixed integer programming: The Vardar-Axios River Basin*. Massachusetts: Massachusetts Institute of Technology. Paper OR 072-78.

* Fernández Lechón, R., Soto, M. D., Garcillán, J. J. (1990). Optimización multicriterio en el contexto de la programación matemática. *Anales de estudios económicos y empresariales*, *5*, 147–160, ISSN 0213-7569.

* Hansen, B. (1996). *Fuzzy logic and linear programming find optimal solutions for meteorological problems*, term paper for Fuzzy Logic course. Nova Scotia: Technical University of Nova Scotia. Retrieved February 3, 2010, from http://www.webindia123.com/career/studyabroad/Canada/details.asp?uname=Technical+University+Nova+Scotia

Kuhn, K., Tucker, A. (1950). Non-linear programming. In: Jerzy Neyman (Ed.), *Proceedings of the second Berkeley symposium on mathematical statistics and probability* (pp. 481–492). Berkeley: University of California Press.

Lee, S. (1972). *Goal programming for decision analysis*. Philadelphia: Auerbach Publishers.

Malczewski, J. (1999). *GIS and multicriteria decision analysis*. New York: Wiley.

* Nijkamp, P., & Spronk, J. (1978). *Interactive multiple goal programming*, *Research Memorandum 3*. Amsterdam: Department of Economics, Free University Amsterdam.

* Schrage, L. (1984). *Linear, integer and quadratic programming with Lind* (3rd ed.). San Francisco: Scientific Press.

Steuer, R. (1986). *Multiple criteria optimization: Theory, computation and application*. New York: Wiley.

Weingartner, H. (1966). Capital budgeting of interrelated projects: Survey and synthesis. *Management Science, XII*(7), 485–516.

---

* Indicates suggested reading material not mentioned in text.

Chapter 5
Features in Formulating and Solving Decision Problems – Sensitivity Analysis

Abstract This chapter starts with the formulation of a case containing enough information to illustrate the construction of the decision matrix and problem solving. Probably the most valuable feature of this example is the thorough analysis performed when using several objective functions. The idea is to demonstrate the wealth of information that can be extracted from the model and how it can detect if there have been shortcomings in establishing the mathematical model. It also reveals how the information provided can help the DM in making clear that some concepts are worth reviewing in the light of information that cast a doubt about early stage concepts. It finalizes with analysis of how weights assigned to the diverse objectives may affect the solution.

Keywords Decision problems • Supporting the decision maker • Weights importance • Pay-off matrix • Requests from the Decision maker

5.1 Case Study: Selecting Environmental Options for Electrical Generation in a Region

The proposed case illustrates the solving of an actual and complex decision problem. It involves looking for the best solution with three alternatives for electrical generation, subject to five important objectives.

Impacts produced by alternatives are studied, and the theme expands to involve other factors, for instance, assuming that all alternatives are feasible from the economic and technical point of view, but normally with different costs, even for the same output to deliver in the three cases. In the analysis of this problem are:

- Social problems, such as noxious gas emissions which can affect human health; consequently it is of capital importance that said emissions be as low as possible (minimization criterion).

N. Munier, *A Strategy for Using Multicriteria Analysis in Decision-Making: A Guide for Simple and Complex Environmental Projects*, DOI 10.1007/978-94-007-1512-7_5,

- Economic problems, since large differences in costs may exist not only for powerhouse construction but also for the ancillary systems (minimization criterion).
- Environmental problems, because of norms and regulations for air contamination that must be honoured (minimizing criterion).
- Spatial problems, such as smoke transported by winds, which deposits soot and other contaminants in areas located far away from the powerhouse that spewed it. The terrible Chernobyl accident is a clear demonstration of this type of problem, since the radioactive particles reached places like Sweden very far away from the Ukrainian plant (maximizing the safety criterion).

To all of this it is necessary to include subjective values which are those from public opinion regarding acceptance or degree of acceptance of the installation of such a plant in an area, or regarding visual contamination, etc.(maximizing and minimizing criteria).

5.1.1 Objectives

The construction of one, two, or three powerhouses in three different places already identified. Each plant has its own characteristics since they operate based on different technical procedures aimed at decreasing environmental pollution, but all of them with the purpose of achieving the following five objectives:

Objective: ***Maximization of electrical energy output*** (Za)
Generate more energy adopting new and distinct technologies, albeit not necessarily the same technology applies to all sites. That is, it is possible to build one, two or three plants and each one can work with the same or different technology.

Objective: ***Minimization of air contamination*** (Zb)
Decrease contamination significantly – with respect to conventional schemes – of gases like CO_2, one of those responsible for global warming, in order to comply with the directives issued by the European Union.

Objective: ***Maximization of rural electrification*** (Zc)
Increment electrical energy supply to rural population, measured in km^2. The size of this area is related with each already pre-selected location, as a function of soil characteristics, population density and industries in the area.

Objective: ***Maximization of job generation*** (Zd)
Generate work opportunities in the region's energy generation industry. Work opportunities relate to the proposed methodology since some of them require more personnel than others in construction and operation, and direct, indirect and induced work positions are considered.

Objective: ***Minimization of purchased energy*** (Ze)
Reduce energy extracted from the national electric grid. The reason is strictly political since it aims at minimizing the dependency of the region regarding electrical energy supplied by the national grid, and since it is argued that it is expensive energy. The grid energy has no direct relationship with locations for the proposed electricity generation schemes in the region, however, it indirectly relates to them, considering the variable construction costs of laying high voltage lines from the grid.

5.1.2 Alternatives

Feasibility studies show three possible alternatives, which are:

(A) Identified as X_1, consists of a thermal plant with fluidized bed boilers.[1] The construction of this type of plant can take place in any of the different locations.
(B) Identified as X_2, utilizes gas turbines fed with liquid fossil fuel. The construction of this type of plant can take place in any of the different locations.
(C) Identified as X_3, consists of an ordinary thermal plant burning pulverized coal, but applying the most innovative technique available nowadays (and without background history, except for a similar plant under construction in the U.S.A.). It aims at elimination of atmospheric contamination by gases spewed by the stack, by sending them to natural deep caves for storage. Not only would the emissions be completely removed but, because of the gigantic CO_2 deposit formed, there will be gas available that could be used by industrial processes in the future (for instance for carbonated drinks, refrigeration units, industrial cleaning, industrial organic syntheses, etc.). This type of plant can only be located in one of the three pre-selected locations, since it is the only one that possesses the adequate geologic characteristics in its underground.

5.1.3 Determination and Evaluation of Impacts for Each Alternative

Considering the characteristics of this project and its relationship with the environment, the first task is to determine the positive and negative impacts that they can produce. See Tables 5.1–5.3 for each alternative.

[1] In fluidized bed boilers the SO_2, from the sulphur present in the fuel, combines with limestone and transforms into gypsum, which is collected with the ashes. This system effectively eliminates the SO_2, NOx, and dust and generates an additional benefit, which is the production of commercial quality gypsum.

Table 5.1 Impacts produced by alternative X_1 using fluidized bed boilers

Type of impact	Impact scope	Degree of mitigation	Contamination value
Primary positive	There will be employment generation in an industry with a high multiplier effect	There is a strong reduction in SO_2	Production, mg/m^3 Permissible: mg/m^3
Primary positive	There will be an additional commercial effect because the sale of gypsum produced because the fluidized bed		Annual amount saleable Euros
Primary negative	Produces atmospheric air contamination with SO_2, NOx and S_2H	With the new electrostatic filters contamination will be just below the maximum admissible threshold, except for NOx. Studies are actually being performed to improve this effect and bring it to acceptable limits	Production mg/m^3 Permissible: mg/m^3
Primary negative ↓	Carbon dioxide contributes to **greenhouse effect**		
Secondary negative ↓	The greenhouse effect produces an **increase in the planet temperature**		
Tertiary negative ↓	The increase in temperature originates **flooding** in many parts of the world		
Quaternary negative	**Flooding leaves hundreds of thousands of people homeless**, and produces the disappearing of lowlands such as the Maldives Islands, Bangladesh and mangroves in the Pacific Ocean		
Primary negative ↓	**SO_2 and S_2H** affect human health producing irritation and lung ailments		
Secondary negative ↓	Produces **acid rain**		
Tertiary negative	**Fish death** in rivers because of acid rain		History of similar impact
Secondary negative	Carbon dioxide produces alteration of pH in rivers and lakes through carbonic acid		History of similar impact

(continued)

Table 5.1 (continued)

Type of impact	Impact scope	Degree of mitigation	Contamination value
Possibility of taking measurements (positive)	Easy to extract samples to measure concentration		
Accumulated negative			History of damages
Is it possible to mitigate?		Yes, but it is necessary to perform studies to learn which is the best system	
Risks	Related with the characteristics of emitted gases		

Table 5.2 Impacts produced by alternative X_2 using gas turbines

Type of impact	Impact scope	Degree of mitigation	Contamination value
Primary positive	There will be employment generation in an industry with a high multiplier effect	Complete combustion	Economic multiplier effect:
Primary negative	**SO_2 and S_2H** affect human health producing irritation and lung ailments	Better combustion in gas turbines reduces very effectively atmospheric contamination	Production, mg/m^3 Permissible: mg/m^3
Primary negative ↓	Affects human health	There is no experience in this respect	Expected percentage cases of lung ailments %
Secondary negative ↓	Produces **acid rain** although possibly in a greater degree than the other systems		Production mg/m^3
Tertiary negative	**Fish death** in rivers because acid rain		History
Possibility of taking measurements (positive)	Easy to extract samples to measure concentration		
Accumulated, negative	On top of health risk it will generate acid rain especially because SO_2		History of damages
Is it possible to mitigate?		It is rather difficult because instalment of exhaust filters produces loses in the turbine output	History of mitigation in similar projects
Risks	Unknown, but studies in similar plants already built some years ago and without the benefit of the advanced filters available nowadays, estimate that there will not be high level harmful effects		

Table 5.3 Impacts produced by alternative X_3 using underground gas storage

Type of impact	Impact scope	Degree of mitigation	Contamination value
Positive	There will be employment generation in an industry with a high multiplier effect. On the other hand this is an experimental technology, and if implemented in the region could place the country's industry in an unforeseeable level of sophistication, especially from the commercial point of view	No necessary	
Adverse	There are no adverse effects in the atmosphere		
Secondary	It is essential to carry out considerable infrastructure work. On the other hand it is necessary to deduct from the generated energy that needed to drive the electric pumps to force the gas underground		
Possibility of taking measurements (positive)	Absolutely		
Risks	As a new technology there are many unknowns and large risks involved because for instance it is not known how the underground storage will work Secondly it is not fully ascertained what the commercial future is for the stored gas, which of course must be purified before it is utilized		

Notice that:

- '*Type of impact*' column indicates which criterion must be maximized or minimized.
- '*Impact scope*' column. In knowing the scope of each impact the corresponding criteria can be developed.
- '*Degree of mitigation*' column indicates effects and measures that can be taken to mitigate impacts.
- '*Contamination value*' column. Scores or contributions for each alternative and criterion can be those of contamination values.

Once the impacts are known it is possible to establish what criteria are needed to evaluate them, at least from the environmental point of view, and of course, it is also necessary to add other criteria related with other issues, therefore this analysis must be repeated in the social and economic fields.

5.1.4 *Working with Different Objectives*

For the sake of simplicity and because the purpose here is to analyze different objectives, those criteria such as investment costs, costs per MW, investment performance, etc., as well as the qualitative data, are not considered, and we will focus only on the compliance of the five objective functions. Of course, adding the above-mentioned values does not alter the analysis, but naturally, will alter the result if incorporated into the matrix.

As seen, there are very diverse objectives, some as maximizing and others as minimizing functions, and a few clearly conflicting, such as Za and Zb, but all of them expressed as linear functions of the three unknowns, which represent the three alternatives, X_1, X_2 and X_3. The problem assumes that all technical and subjective scores associated to each alternative are determined. Thus, the problem can be mathematically set out as follows:

Maximize $Za = 250X_1 + 321X_2 + 220X_3$ (The scores or coefficients indicate the design output for each powerhouse). These are technical coefficients derived from the physical and thermodynamics characteristics of each plant, and the number of generation units, i.e. steam or gas turbines as prime movers of electricity generators.

Minimize $Zb = 234X_1 + 418X_2$ (The scores or coefficients indicate the production of CO_2, in tons per MW). This value does not exist for a powerhouse represented by unknown X_3 because the gas produced is stored underground and consequently, not released to the atmosphere.

Maximize $Zc = 447X_1 + 823X_2 + 98X_3$ (The scores or coefficients indicate km^2 of rural territory which will benefit from laying electrical cables for rural electrification, with energy generated by each powerhouse).

Maximize $Zd = 1652X_1 + 2234X_2 + 2381X_3$ (The scores or coefficients indicate the direct and indirect jobs which can be created depending on the type of powerhouse, and during construction and operation).

Minimize $Ze = 124X_1 + 356X_2 + 87X_3$ (The scores or coefficients indicate MW that can be extracted from the national grid).

5.1.4.1 Twofold Role (Objectives and Criteria)

We have seen that besides the alternatives there are two basic elements in a decision problem: objective/s, and criteria. These two mathematical expressions are very similar since both are linear equations and use the same set of variables or alternatives, and consequently, can be utilized indistinctively as objective functions or as criteria.

Kumar et al. (2003), hypothesize that the objective function and restrictions are of the same nature and that the distinction between them is then gradual and not abrupt. This concept is then crucial in this work since it dilutes the difference between objective functions and attributes, as is maintained in this book when developing the SIMUS variant.

However, their meaning and purpose is clearly different according to its utilization. An objective is a *goal*, something that we wish to achieve. A criterion is a *target*, which establishes an action, a way or method to reach that goal or make it achievable. There is another fundamental difference since a goal for a project or plan is generally indefinite or imprecise, such as '*Minimize the ecological impact*' (notice that this is a general expression, it does not say how to attain it, or by using what means, although sometimes a completion time is specified, for instance five years). On the other hand, criteria or targets employed to achieve that goal are definite and precise and even delimited, such as '*Minimize water consumption*' (and establishes a limit for this consumption), or '*Maximize forested area*' (and perhaps fixes the minimum number of hectares, species to be used, etc.). These restrictive expressions can be thought as means to attain an objective.

Because this double role, from now on and to avoid confusion, an expression is called '*objective function*' when is used as the objective or goal of the problem, while it is called '*target*' when utilized as a criterion.

5.1.4.2 Independent Terms

The independent terms, thresholds, or Right Hand Side (RHS) values are the limits imposed to targets. Thus, for the first objective when working as a target – *Maximization of electrical energy output* – it is established that there must be at least 245 MW generated, and the sign'$\geq$' will be used. What is the reason for this value? It derives from economies of scale, since it is not worth building power plants that will generate less output.

For the second objective when working as a target – *Minimization of air contamination, measured in CO_2 production* – an upper limit of 250 tons has been set up, and hence the sign '$\leq$' will be utilized. What is the reason for this value? Because it is the maximum allowable amount for matching in three years time, a goal established by the Kyoto Protocol[2] and the European Union Norms to reduce contamination. Thus, it expresses as a target that contamination must reach a maximum of 250 tons of CO_2 or less.

For the third objective when working as a target – *Maximization of rural electrification* – a threshold of 300 km^2 for rural areas is created, and it is a lower limit, established considering rural population density, therefore, the result must exceed that limit.

[2] Refers to the Kyoto Protocol on Climate Change, an international agreement aimed at reducing emissions of gases causing global warming.

For the fourth objective when working as a target – *Maximization of job generation* – a value of 1,500 signals the minimum threshold for creating new employment, and includes construction, operation and maintenance jobs as well as indirect and induced jobs. Thus, the result must exceed this limit.

For the fifth objective working as a target – *Minimization of purchased energy* – the plan establishes as a limit that the maximum extraction from the nation's grid must be 175 MW. This limit may be subject to different reasons, and local and political considerations are probably included in fixing this limit.

As seen, the mathematical arrangement includes all the technical aspects as well as the expressions of interest (maximization or minimization), for the different objectives. Table 5.4 condenses this data.

Table 5.4 Decision table – summary of objective functions scores and independent terms

Objective functions		Alternatives			Action	Operator	Independent terms or RHS values
		X_1	X_2	X_3			B
Za	Electrical energy output	**250**	**321**	**220**	Max	≥	**245**
Zb	Air contamination (CO_2 production)	**234**	**418**		Min	≤	**250**
Zc	Rural electrification	**447**	**823**	**98**	Max	≥	**300**
Zd	Job generation	**1,652**	**2,234**	**2,381**	Max	≥	**1,500**
Ze	Purchased energy	**124**	**356**	**87**	Min	≤	**175**

5.1.5 Construction of the Decision Matrix

An Excel spreadsheet is used to build the decision matrix (it is not shown here because in reality it adds nothing substantial). However, it is convenient to mention that in this last table there is a restriction or target (Z'a) added, which establishes a maximum limit for electric generation of 450 MW. What is the reason for this value? It comes from the amount of available funds to finance the project. We can use the Solver® or any other Linear Programming software, such as LINDO®, MATLAB®, etc., to get the solution for the problem proposed.

As mentioned in our previous discussion an objective function is chosen, and any of the objectives will do the job. We can start with Za, with the maximization action, while Zb, Zc, Zd and Ze will remain as targets in the decision matrix.

5.1.5.1 Solving for Za

Za is removed from the decision matrix and used as objective function, the Solver is run and results are obtained; the shaded row in Table 5.5 shows the result. We get an optimal solution from the point of view of 'Maximization *of electrical energy output*'.

Table 5.5 Computed values for alternatives for each objective function

Alternatives		X_1	X_2	X_3
Objective functions	Objective description	Values for the alternatives for each project		
Za	**Maximization of electrical energy output**	**0.3347**	**0**	**1.5344**
	Ranking:	Second		**First**
Zb	Maximization (CO_2 production)	**0.3347**	**0**	**1.5344**
	Ranking:	Second		**First**
Zc	Maximization of rural electrification	**1.0684**	**0**	**0.4888**
	Ranking:	**First**		Second
Zd	Maximization of job generation	**0.3347**	**0**	**1.5344**
	Ranking:	Second		**First**
Ze	Minimization of purchased energy	**0.5687**	**0**	**0.4674**
	Ranking:	**First**		Second

Notice that in this case, the model shows a preference of alternative X_3 over $X_{1,}$ while alternative X_2 is not considered. Alternative X_3 gets a score of 1.5344 (First or preferred choice), while alternative X_1 gets only 0.3347 (Second choice). Thus, the ranking is X_3–X_1.

Multiplying each score from Table 5.5 by the respective coefficients of variables X_j, from Table 5.4 and adding up these results, gets the value of each objective. Thus, the output for objective function Za is 421 MW (underlined in Table 5.6). Notice that this value exceeds, as required, the minimum limit of 245 MW, and at the same time it is below the maximum limit of 450 MW. If the found values for the alternatives are replaced in the other equations corresponding to the different Z_j (targets), we can get their respective outputs as a function of Za, which are depicted in Table 5.6.

Table 5.6 Pay-off vector corresponding to objective function Za

	Za	Zb	Zc	Zd	Ze
Za	**<u>421</u>**	**78**	**300**	**4206**	**175**

These outputs are provided by the Solver, and they can be verified as follows:

$$Za = 250 \times 0.3347 + 220 \times 1.5344 \quad = 421$$
$$Zb = 234 \times 0.3347 \quad = 78$$
$$Zc = 447 \times 0.3347 + 98 \times 1.5344 \quad = 300$$
$$Zd = 1652 \times 0.3347 + 2{,}381 \times 1.5344 = 4206$$
$$Ze = 124 \times 0.3347 + 87 \times 1.5344 \quad = 175$$

All of these objectives comply with the respective criteria since:

Zb = 78, is less than the maximum which was 250 ton CO_2.
Zc = 300, is larger than the minimum which was 250 km^2.
Zd = 4,206, is larger than the minimum which was 1,500 jobs.
Ze = 175, is equal to the maximum which was 175 MW.

5.1.5.2 Solving for Zb

Za is restored in the decision matrix as a target, and we proceed with the second objective that is Zb – Air contamination with CO_2 production – by removing it from the decision matrix and used as objective function, but now with the minimization action. Table 5.7 displays the result.

Table 5.7 Pay-off vector corresponding to objective function Zb

	Za	Zb	Zc	Zd	Ze
Zb	**421**	**78**	**300**	**4206**	**175**

Coincidentally, identical values appear regarding Za, although the objective functions are completely different and even with opposed actions.

By repeating the procedure for objective functions Zc, Zd and Zd, it is then possible to build the pay-off matrix of all objectives (Table 5.8).

This is the first working tool for the DM.

Table 5.8 Pay-off matrix for each subjective function

When using these objective functions ↓	Za	Zb	Zc	Zd	Ze
	These are the output values for targets ⟷				
Za	**421**	**78**	**300**	**4206**	**175**
Zb	**421**	**78**	**300**	**4206**	**175**
Zc	**375**	**250**	**525**	**2929**	**175**
Zd	**421**	**78**	**300**	**4206**	**175**
Ze	**245**	**133**	**300**	**2052**	**111**

5.1.6 *Informing the Decision Maker – Analysis of Results – The Pay-off Matrix – Its Usefulness to the Decision Maker*

Now we have for each objective function the different outputs obtained for the targets and then it is possible to study the reasons for reaching these results. This examination is performed observing simultaneously the corresponding values for Tables 5.4, 5.5 and 5.8.

Remember that:

Table 5.4 shows <u>scores</u> or <u>coefficients</u> assigned to each objective of the project, and they are quantitative (as exemplified here). If we had for instance a qualitative objective such as '*People opinion about the project*', then there would be qualitative or subjective values, coming from personal appreciation or from a survey.

Table 5.5 shows <u>values of alternatives</u> and their ranking according to each objective function. Table 5.8 shows the payoff matrix that is the <u>output values for targets</u> corresponding to each objective function. We have now the results, but not *their analysis*, that is, there is interest in knowing the *reasons for these results* and conclusions we can possibly extract from them, and that is what follows.

5.1.6.1 Analysis of Results

Objective function Za, with a minimum required generation value (threshold) of 245 MW (Table 5.4), gets now 421 MW (row **Za**, column Za, Table 5.8) and for the targets we have:

Target Zb:
According to Table 5.4 the maximum admissible value for CO_2 is 250 tons (column B). However, when the objective function calls for maximization of the installed capacity, this contamination decreases abruptly to 78 tons, (row **Za**, column Zb, Table 5.8).

This appears counterintuitive, since it is reasonable to expect that increasing energy generation will produce a raise in CO_2 production. The explanation lies in Table 5.5 where it can be seen that, in accordance with the Za objective function, project X_2 is not to be executed. Why not?

Because in accordance with Table 5.4 powerhouse X_2 has the maximum contamination value (418), and since this is a strong restriction it produces the elimination of the alternative.

Target Zc:
The requirement is to have a minimum of 300 km^2 (Table 5.4), and it does not change as shown in row **Za**, column Zc, Table 5.8.

Why not?

Because even not building plant X_2, the required total area of 300 km^2 can be more than supplied by the other two options.

Target Zd:
It increases drastically employment creation (from 1,500 to 4,206) (row **Za**, column Zd, Table 5.8).
Why?

Because even when X_2 is the second option regarding number of jobs created (2,234) (Table 5.4), job opportunities corresponding to X_1 y X_3, more than compensate for no execution of X_2.

But, why are more jobs than the minimum required generated?

Because of the ratio between jobs and output generated, which is a ratio of technical coefficients.

Observe (Table 5.4), that for X_1 this ratio is 1,652/250 = 6.61 jobs/MW; for X_2 it is 2,234/ 321 = 6.95, but for X_3 it is 2,381/220 = 10.82. Consequently, when increasing the generation of X_3 because of no execution of X_2, it strongly increases employment creation.

Target Ze:
The maximum planned extraction of 175 MW (row **Za**, column Ze, Table 5.8) holds, which means that the Za objective function has no influence on energy extraction from the grid, which seems logical, since values corresponding to Za in Table 5.4 indicate capacities of each plant and not the needs of each zone. If another restriction specifying a minimum amount of energy to be produced had been added, then

probably that would influence the quantity of energy to import. *In reality when this conclusion is presented to the DM, he/she could discover that the formulation of the problem was faulty – due perhaps to him/herself and staff omission – because at that moment he/she could have realized (as would be logical), that it was necessary to establish a forecast of demand and make sure that it was honoured.*

That is, this could be as good as to say that the model was not set out correctly, that it is incomplete. *However, the model by itself is unable to detect this fault, which manifests only when the information it provides is analyzed.* In this case the problem should be corrected and the process of analysis repeated in an interactive feed-back and working relationship between the DM and the mathematical model.

Objective function Zb, with a generation value of 250 ton CO_2 now produces only 78 ton, and targets:

Target Za:
There is a large increase in energy production (from 245 to 421 MW) (row **Zb**, column Za, Table 5.8).

Why? The logic says that if we want to decrease contamination energy, production should also decrease. The reason again is the non-execution of X_2, because it is the plant that produces maximum contamination (418 tons) (Table 5.4).

Target Zc:
Holds constant and adjusted to the design value (row **Zb,** column Zc, Table 5.8).

Target Zd:
Sharply increases to 4,206 jobs (row **Zb**, column Zd, Table 5.8) for the same reasons as already explained.

Target Ze:
Holds constant and adjusted to the design value of 175 MW (row **Zb**, column Ze) for the same reason mentioned when Za was analyzed.

Objective function Zc, with a minimum area value of 300 km^2. When maximizing this area goes up to 525 km^2 (row **Zc**, column Zc, Table 5.8).

Targets take these values:

Target Za:
Increases electrical generation from 245 to 375 MW (row **Zc**, column Za, Table 5.8). This is rational, since the expansion of the electric service from 300 to 525 km^2 needs an increase in power generation.

Target Zb:
We already know that the number of hectares increases from 300 to 525. This translates in the necessity of expanding the electric network and the generation of more electric energy, which in turn will produce more contamination. However, the contamination remains the same at 250 (row **Zc**, column Zb, Table 3.5).

How can it be explained?

Because the increment in contamination due to an electric generation increase to 375 MW, is produced by alternatives X_1 and X_3 since alternative X_2 does not appear in the solution (row **Zc**, column X_2, Table 5.5). Considering that X_3 produces zero contamination, it is obvious that total contamination can remain the same, since a large part of it goes underground.

Target Zd:
The number of jobs increases to 2,929 (row **Zc,** column Zd, Table 5.8).

This is explained, because of the large increment (75%), in the area served, which is now 525 km² in lieu of 300 km² , involving the electrical grid expansion and maintenance, transformers construction and installation, personal transportation, more personal working in the powerhouses, more clerical work, etc.

Target Ze:
It holds the maximum amount of extracted energy, i.e., 175 MW (row **Zc,** column Ze, Table 5.8), because the necessary increase of electric power comes from alternatives X_1 and X_3, so there is no need to import more energy.

Objective function Zd, with 1,500 work positions. If the objective of creating more jobs is maximized, the results show a total of 4,206 jobs (row **Zd,** column Zd, Table 5.8). This sharp increase is due to the fact that when Zd is the objective function alternative X_3 is 4.53 times more important than alternative X_1, and also because X_3 has the largest value for numbers of jobs created. Targets Za and Zb have the same results as in objective functions Za and Zb, and then their analysis is not repeated here.

Objective function Ze, with a value of 175 MW for extraction from the grid, the extraction decreases to 111 MW (row **Ze**, column Ze, Table 5.8), and targets are:

Target Za:
Electric generation is almost distributed in equal parts involving powerhouses X_1 and X_3 (row **Ze**, column X_1 and **Ze**, X_3, Table 5.5), with a total of 245 MW (row **Ze,** column Za, Table 5.8).

Target Zb:
Decreases from 250 to 133 tons of CO_2 production. (row **Ze**, column Zb, Table 5.8).

Why?

Because plant X_2 is not considered, and this is the plant producing the largest contamination.

Target Zc:
Holds the specified limit of 300 km² (row **Ze**, column Zc, Table 5.8), because the same reasons explained for objective functions Za and Zb.

Target Zd:
This is the alternative which produces the least increase in jobs creation (row **Ze**, column Zd, Table 5.8).

Why?

Because of decreased energy extracted from the grid and consequently there are fewer customers.

At the end of this analysis, the question is which is the best ranking of powerhouses? See discussion in Sect 5.7, but without a doubt it is:

$$\boxed{X_3 - X_1}$$

Because for objective functions:

$$\mathbf{Za}, \text{the result is}: \mathbf{X_3} >> \mathbf{X_1}$$
$$\mathbf{Zb}, \text{the result is}: \mathbf{X_3} >> \mathbf{X_1}$$
$$\text{Zc, the result is}: X_1 >> X_3$$
$$\mathbf{Zd}, \text{the result is}: \mathbf{X_3} >> \mathbf{X_1}$$
$$\text{Ze, the result is}: X_1 > X_3 \text{or } X_1 \approx X_3$$

This was an extensive and complicated analysis, and many people will probably be reluctant to try to apply it in another actual case, however, this was made difficult on purpose to show the potential of LP in analyzing different scenarios. Of course, it can be argued that there is a lot of information that probably the decision maker cannot digest, and this is true, but again, the purpose was to show the kind and variety of information the LP model can furnish. This example demonstrates that LP can give the decision maker a ranking of alternatives, all duly supported, and the reasons for each variation are explained in detail. The decision maker has then an objective tool that is transparent and that can document his/her decision.

5.2 Informing the Decision Maker – Importance of Different Objectives, and Analysis of Their Influence on Alternatives Ranking

A question that surely any DM puts forward in the initial stages of the study of the project is in what measure an objective is more important than another. This is a question whose answer is not easy considering the different units between them; i.e. some objectives might be expressed in Euros, others in concentrations of DBO_5, others in units such as km, or in m^3, etc., and besides there might be some objectives without any dimension, such as for instance the discomfort for the noise produced by wind turbines.

Linear Programming can help in answering this question through a sensitivity analysis which is automatically generated each time a problem is solved, at least in the Solver program. Coming back to the proposed example, it can be verified that objective **Ze** is the most significant of the five objectives. The reason for this assertion can be found in Sect 5.3.

If this assertion is accepted, it is obvious that it is necessary to have a deeper look at the energy extraction from the grid to be able to answer questions like these:

- Why is the purchase of energy so important?
  Without going into a deep analysis it is probable that the analyst's answer is that it is due to the so-called '*Opportunity cost*'[3] of electricity.
- Computation demonstrates (but not shown here), that importing electricity in a quantity larger than foreseen reveals the need to build alternative X_2, and also increase the contribution of X_3.
  Why is alternative X_2 now considered? Possibly because there is a multiplier effect, i.e., having more energy from the grid, there is greater availability and a larger incentive to consumption, more industries, etc., and if the imported quantity holds fixed, that increase is necessary to compensate with X_2.
  The purpose of this last comment is to point out how the model can make the decision maker see *things and aspects that he/she had not perceived or in which he/she was mistaken of perhaps fueled by political reasons*. Due to the complex play of interactions between diverse objectives and targets and criteria, it is many times impossible to have a clear idea of how a decision can affect a project; and from this point of view, LP is invaluable.

The problem just analyzed, albeit relatively complex is, no doubt, incomplete; it would be necessary to define many more concepts as for instance energy demand, its generation costs, benefits derived from this undertaking, etc., but it is evident in this approach that the tool supplies data to make a right decision.

The aspect of energy import for instance, handled as a political banner, probably will produce more harm than benefits, but the 'probably' is not enough; it is necessary to support it with numbers.

Another criterion which would dictate changes in the scheme would be fixing a higher limit of electrical generation, which most probably would provoke the three plants into producing larger capacity; but, of course, that would also bring an increase in CO_2 production, which would perhaps exceed the maximum allowable limit, which is not permitted. As appreciated, with LP the decision maker and stakeholders have information allowing comparisons and adopting the most convenient solution. Naturally, the proposed method does not mean that multiple objectives are optimized, since they are analyzed separately, but what is obtained with this system is a large, unified, scenario of options.

5.3 Determining the Importance or Weight of Each Objective

Until now, it has been assumed that the decision maker or decision entity has not shown any type of preference regarding objectives, that is we have assumed that all have the same weight. Naturally, in case one wishes to do so, it is always possible

[3] This very important economic concept indicates the opportunity that is lost when a resource – in this case the electrical energy which is available for a certain use and is not purchased – is applied to other uses.

to show these preferences by assigning weights to them. It could very well be that there is no preference because the decision maker estimates that all objectives are equally important, since they satisfy varied interests, and because it is very difficult to establish differences, and probably this is a right reasoning. Another motive could be that, in these circumstances, the analysts had advised the DM not to establish preferences since the model (LP) can do that, because it produces a set of *shadow prices*[4] or marginal values for the targets corresponding to the solution for each objective function.

These shadow prices, obtained when running the model with a certain objective function, correspond to the other objectives (targets) that are in the decision matrix. It is worth remembering that this concept of '*weights*' in LP is different from when it is applied in other methods, where generally these are percentage values, and consequently when one of them varies, its variation affects the others. Here the shadow prices have a different meaning since they represent the change in the objective function for a unit variation of the target threshold, and as a consequence, it is also the marginal value of a target, whatever the unit of measure is expressed.

Shadow prices are then interpreted as 'weights' of the different targets, and consequently, it is possible to establish a target hierarchy (or a objectives hierarchy, in this case), in accordance with their respective influence on the objective function. Let's see these shadow prices, and how they are interpreted in each case.

When using **Za** as objective function, it has a value of 421 MW (Table 5.8).

There are two shadow prices corresponding to targets Zc and Ze, whose values are:

$$\begin{aligned} Zc &= -0.207, \\ Ze &= 2.762. \end{aligned}$$

Explanation of shadow price for target Zc: The variation of 10 units in Zc (ten units are considered instead of one to better visualize the variation of the objective function[5]), that is from 300 to 310 km^2, decreases the value of the objective function Za by 2.07 MW, i.e. to 419 MW (Rounding to 2 MW).

Conclusion: When area served increases, there is a smaller share of available energy between more users.

Explanation of shadow price for target Ze: When varying one unit of Ze, that is from 175 to 176 MW, the objective function Za raises by 2.762 MW from 421 to 424 MW (Rounded).

---

[4] Shadow prices values are automatically produced when using the Solver, and are found in the '*Sensitivity*' tab.

[5] This proportionality can take place because we are in linear programming. Consequently, when varying the surface of the area in one unit, from 300 km^2 to 301 KM^2, the objective function will increase by 0.207 MW. Since this is such a small quantity to visualize, a variation of 10 units is adopted, and the area goes then from 300 km^2 to 310 km^2. This procedure would not be valid if criteria were not linear, because then the 0.207 value would be legitimate only in one point and it could not be possible to extrapolate linearly.

Conclusion: It improves the availability of electrical energy when more energy is imported from the grid, a fact that does not need any explanation.

When using **Zb** as the objective function, whose value is 78 tons of CO_2 (row **Zb**, column Zb, Table 5.8), the shadow price for targets Zc and Ze are:

$$Zc = 0.761,$$
$$Ze = -0.858.$$

Explanation of shadow price for target Zc: When Zc varies by 10 units, that is from 300 to 310 km^2, it boosts the value of the objective function Zb by 7.61 tons, i.e., from 78 to 87 tons.

Conclusion: It increases because when there is more consumption, direct and indirect contamination rises.

Shadow price for target Ze: Raising by 10 units the energy extracted, that is from 175 to 185 MW, this increment decreases the value of the objective function Zb by 8.58 tons, i.e. from 78 to 70 tons.

Conclusion: It diminishes because there is an increase in available energy, however, as it is generated at another site, it does not affect the region.

When using **Zc** as objective function, whose value is 525 km^2 of rural area (row **Zc**, column Zc, Table 5.8), the shadow prices corresponding to targets Zb and Zc are:

$$Zb = 1.313,$$
$$Ze = 1.126.$$

Explanation of shadow price for target Zb: Varying by 10 units the $CO_{2,}$ contamination, from 250 to 260 tons, that is relaxing the limit, increments the value of the objective function Zc in 13.13 km^2, i.e., the values goes from 525 to 538 km^2.

Conclusion: It rises because, allowing an increase in the maximum contamination limit, it is possible to serve more users.

Explanation of shadow price for target Ze: Boosting by 10 units the extraction of energy from the grid, the value of the objective function Zc increases by 11.26 km^2, from 525 to 537 km^2.

Conclusion: It rises because of augmentation of the number of users.

When using **Zd** as objective function, whose value is 4,206 jobs (row **Zd**, column Zd, Table 5.8), shadow prices are:

$$Zc = -5.667,$$
$$Ze = 33.751.$$

Explanation of shadow price for target Zc: Varying by 10 units the area served, that is from 300 km^2, there is a decrease from 4,206 to 4,150 jobs.

Conclusion: It decreases because of the increase in the area served, by 10 km², even though small, produces a change in the proportion of the energy matrix. In effect, X_3 is now 1.488 (instead of 1.5344), that is it decreases, while X_1 increases to 0.3673 (from 0.3347). Since the decrease is larger in X_3 when compared with the increase in X_1, it leads to a decrease in jobs.

Explanation of shadow price for target Ze: Varying 10 units the energy extraction, that is from 175 to 185 MW, the value of the objective function Zd increases by 337.51 positions to 4,544 jobs.

Conclusion: The shadow price rises because, in incrementing the availability of energy, more personnel is needed to serve more users.

When using **Ze** as objective function, whose value is 111 MW (row **Ze**, column Ze, Table 5.8), the shadow prices are:

$$\text{Zc} = 0.0749,$$
$$\text{Za} = 0.3621.$$

Explanation of shadow prices for target Zc: Varying by 10 units the area served, that is from 300 to 310 km², the value of the objective function Ze will remain practically constant since it will increase only by 0.479 MW.

Conclusion: It is a small raise because the increase in the area is very small when compared with the original value, which is 10/300 or 3.33%.

Explanation of shadow price for target Za: Varying by 10 units the electricity generation, that is from 245 to 255 MW, the value of the objective function Ze increases by 3.621 MW, that is from 111 to 115 MW.

Conclusion: Logically, if imports rise, the availability of electrical energy increases.

Looking at the above values, it is evident that objective **Ze** has the *greatest influence*; consequently it seems logical that the decision-maker considers this circumstance, especially because, as mentioned before, X_1–X_3 preference is weak. Thus, the decision-maker may request a deeper analysis of the values related to Ze.

5.4 Addressing Requests from the Decision Maker – Feed Backing the Model

Most probably, the DM will have many questions, in which case it would be worthwhile to sample a few of them to see how the model can help in answering them.

Q. What happens if energy import decreases to 100 MW?

A. The answer to this question consists in finding out how the model reacts when energy extraction decreases from 175 to 100 MW, since the DM personally honours the policy of minimizing the import (which also may transpire as political pressure from local politicians). The DM wants also to know if in that case there is modification of the $X_3 - X_1$ energetic matrix. The model is loaded with the new data replacing the 175 MW value by 100 MW, and using Za as objective function, that is to maximize energy.

The result shows:

$$\begin{aligned} Za &= 214, \\ Zb &= 143, \\ Zc &= 300, \\ Zd &= 1675, \\ Ze &= 100. \end{aligned}$$

The corresponding values for X_1 and X_3 powerhouses are:

$$\begin{aligned} X_1 &= 0.6096, \\ X_3 &= 0.2805. \end{aligned}$$

As seen, the model replicated the $\mathbf{X_1 - X_3}$ energetic matrix but modified the ranking. This evidently satisfies the DM since the solution does not change substantially when reducing by 43% the volume of energy purchased.

It seems, by logical thinking, that it would be convenient to go ahead with this reduction, but wait, let us see what the other figures say.

If we compare these new values (The new table with the results of the reduction is not shown) with those of Table 5.8 (which imports 175 MW) it is found that:

- Total energy production decreases from 245 to 214 MW, that is a reduction of 12,6% – Negative aspect,
- Sharp decrease in CO_2 production, from 250 to 143 tons, that is 42.8% – Positive aspect (at country scale, because it doesn't affect the region),
- Area served has not changed – Positive aspect,
- Decrease in the number of jobs since they go from 1,875 to 1,675, that is a 10.7% reduction –Negative aspect.

With these conclusions, the DM perhaps realizes that his/her observance of the scheme of reducing energy purchasing does not look as good as thought, and that in all honesty *the region will be in worse shape because there will be less energy available for its industries and especially because of the decrease in job creation.*

Q. *What is the benefit we get by importing energy*?

A. In viewing the results, it is possible that the DM questions if it is a good idea to import energy. Perhaps he/she thinks that it is not beneficial to insist in pursuing pre-electoral banners, and then changes his/her position and demand to know what would happen if instead of importing 175 MW, this amount increases to 200 MW.

The model is run again loaded with the new threshold. The result surprises because, against all political insistence on energy independence, it is revealed with crystal clear evidence that the region will benefit by importing that additional 15% more, according to this result:

$$
\begin{aligned}
Za &= 450\ \text{MW},\\
Zb &= 65\ \text{Tons.}\\
Zc &= 300\ \text{Km}^2,\\
Zd &= 4631\ \text{Jobs},\\
Ze &= 200\ \text{MW}.
\end{aligned}
$$

That is:

- The available energy for the region will reach the highest value of 450 MW, obviously because of the greater import,
- Contamination will decrease to its lowest limit, 65 tons, because part of it is produced elsewhere,
- The area served holds constant at 300 km^2, which is the design value,
- Jobs, against popular belief, greatly increase to 4,531 positions.

However, the energy matrix is now different, since the change in importing makes it necessary to build a powerhouse X_2, and the ranking is now $X_3 - X_1 - X_2$. Frankly, this result is surprising, because if more energy is purchased why is it necessary to build an additional powerhouse? The detailed analysis of the decision matrix and Tables 5.4, 5.5, and 5.8 clearly justify this result.

Q. Considering the importance of objective Ze, is it possible to use any other methodology to back up this result?

A. Certainly, the entropy method allows us to determine which target has the larger influence or importance. It is based on computing the discrepancies between the values of the different alternatives regarding a certain objective, that is, the larger the disparity, the lower the entropy value, and the more important the criterion. This computation has been performed for this example, and it substantiates the selection of objective Ze as the most important.

Q. How stable is the solution found?

A. The DM can argue, and with reason, that he/she needs to know the stability of the solution found by $X_3 - X_1$ since there are subjective factors which can change, and in turn change some of the quantitative factors.
For that reason the DM needs to know between which limits the important parameters such as thresholds may vary, without modifying the solution found. This is part of the sensitivity analysis, for which LP produces three types of data, as follows:

Response: Set of values that indicate how far from the established requirements are the solutions found.

Sensitivity: It expresses the weight or relative importance of each target. As seen, this is done using the shadow prices, thus determining how much the objective function varies for a unit variation of one target or from various targets changing simultaneously.

Limits: Supply information expressing between which limits the values found for each variable can vary without altering the result.

As we can see, this information has a fundamental importance for the DM, and in fact part of it was used when the battery of solutions was submitted to the DM. Naturally there are many other questions that the DM might formulate, but this brief description sheds some light on the potential benefits of this type of analysis.

5.5 Regarding Criteria Pondering

In general there could be various cases regarding DM preferences once the criteria are identified. They are:

(1) The DM wants to assign a relative weight to each criterion, based in his/her personal opinion, experience or needs, as well as in discussion with his/her staff. Weights can also be computed using a system such as AHP, commented on in Sect 3.2.4.
(2) The DM does not want or can't assign weights to criteria because is not confident about which values to use, and prefers giving the same weight to them all. According to the model adopted for decision-making, this procedure poses the risk of overestimating some criteria and underestimating others.
(3) The DM does not want or cannot assign weights to criteria, because considers that due to the different units of measure, it is very difficult to assess their relative importance to the project.

Criteria can be restrictive, that is limiting the scope of decisions, such as establishing an upper limit on the amount allowed for expenses. Criteria may express requirements for individuals to comply with certain conditions, for instance establishing a lower limit for quantity of water available per person, but at the same time establishing an upper limit (in the same criterion but in another row) to discourage water squandering.

Assigning weights to criteria in LP does not alter the procedure, and regarding restrictions and requirements (thresholds), these are characteristics of the Simplex method and constitute part of the theory on what it is based. Let us see instead what happens with the objectives, which in turn have two variations:

– A common unit of measure for all objectives,
– A different unit of measure for each objective.

5.6 Weighting the Objective Functions – Shadow Prices and Their Analysis

Table 5.9 replicates Table 5.4, which is the initial table for objectives' coefficients.

Table 5.9 Decision table – Summary of scores for objective functions and independent terms values

Objectives related with:		Alternatives X_1	X_2	X_3	Action	Operator	Independent terms B
Za	Electrical energy output	**250**	**321**	**220**	Max	≥	**245**
Zb	Air contamination (CO_2 production)	**234**	**418**		Min	≤	**250**
Zc	Rural electrification	**447**	**823**	**98**	Max	≥	**300**
Zd	Job generation	**1,652**	**2,234**	**2,381**	Max	≥	**1,500**
Ze	Purchased energy	**124**	**356**	**87**	Min	≤	**175**

Table 5.10 shows values obtained for alternatives, but with weights assigned to each objective. Note that it appears that the DM wants to favor two main areas, (objective Zb, environment, and Ze, energy import) because both are assigned the highest weight, 25%.

Table 5.10 Computed values for alternatives for each weighted objective function

Alternatives / Objectives functions	Criteria weights	Objective description	X_1	X_2	X_3
			Values corresponding to alternatives for each objective function		
Za	0.2	Maximization of electrical energy output	**0.3347**	**0**	**1.5344**
		Ranking:	Second		**First**
Zb	0.25	Minimization of air contamination (CO_2 production)	**0.3347**	**0**	**1.5344**
		Ranking:	Second		**First**
Zc	0.15	Maximization of rural electrification	**1.0684**	**0**	**0.4888**
		Ranking:	**First**		Second
Zd	0.15	Maximization of job generation	**0.3347**	**0**	**1.5344**
		Ranking:	Second		**First**
Ze	0.25	Minimization of purchased energy	**0.5687**	**0**	**0.4674**
		Ranking:	**First**		Second

As can be seen when comparing Tables 5.5 (not weighted) and 5.10 (weighted), the selections have not changed and this is really surprising, however, it can be interpreted that the solution found is quite stable since it does not change when applying these weights. Of course, this is not a rule, since it is expected that, when applying other weights, the solution will vary.

Table 5.11 shows shadow prices without weight and with weights '*'. Notice that:

Table 5.11 Shadow prices with different objective functions (weighted)

	Za	Zb	Zc	Zd	Ze	
Za			**−0.207**		**2.762**	Max
			***−0.372**		***1.390**	
Zb			**0.761**		**−0.858**	Min
			***2.548**		***−0.802**	
Zc		**1.313**			**1.126**	Max
		***0.393**			***0.315**	
Zd			**−5.66**		**33.751**	Max
			***−0.931**		***1.551**	
Ze	**0.361**		**0.074**			Min
	***0.719**		***0.268**			

- Shadow prices coincide in targets affected,
- Plus and minus signs coincide in all values,

  Of particular interest is the shadow price corresponding to objective function Zd and target Ze. When no weights are used there is a very large influence of target Ze since a unit variation provokes an increase of 33.751 in the objective function, while when using weights that improvement is drastically reduced (Table 5.11). Why?

  Because when no weights were used all criteria have the same importance, but when a weight of 0.15 is applied to Zd and higher weights to others, obviously its influences decreases.

5.6.1 Disadvantage of Weight Estimate by the Decision Maker

Suppose that the DM decides to make a subjective assessment of the objectives and criteria weights, although logically based on aspects that he/she and staff consider suitable, and sometimes aided by the Delphi method (Sect. 2.1). However, there is an aspect that normally is not considered and manifests when the DM estimates the weights at his/her best knowledge and expertise, for he/she may not know if the values assigned match the technical restrictions.

To illustrate this concept consider another project, and assign a weight of 20% to a criterion like project financing, and successively allocate percentages to other criteria to complete the unit. It could be that, if this financing percentage is small, when multiplied by the alternatives score in the '*Financing*' criterion, it generates an amount of money which is not large enough to cover 100% of this objective, and besides it can have an indirect effect on others. Again, this is not known by the DM when assigning these weights, however the Simplex method can detect it, because the Solver would indicate that under these conditions there is no solution for the problem, and clearly will show which restriction is causing the difficulty, allowing for its revision. In a case like this, the Solver ends a message that says:

Solver has not found a feasible solution

Why? Probably, because the new weighting of some of the mandatory requirements have not been honored.

Coming back to the example depicted in the last section, it is easy to determine that the restriction that does not comply is that belonging to objective Zc (*Maximization of rural electrification*), because the weighting in Zb and Zd changes the energy matrix, and this affects Zc because it cannot comply with the requirement that it must serve at least 300 km^2 of rural areas. As can be seen in LP, the model is charged with the task of informing us if it is possible or not to go ahead with those weights.

Suppose now that the DM decides to run the AHP model to determine criteria weights and then apply them to the LP process. Calculating and applying these new weights, and using Za as objective function, the model shows an exclusive preference for alternative X_3, that is, as the unique alternative, and then varying the energy matrix. These two instances illustrate how the assigning of weights can change the solution.

5.7 Selection and Alternatives Ranking

Let's see now the ranking of alternatives in accordance with the subjective function (See Table 5.5):

- From the point of view of objective function Za, that is '*Maximization of electrical energy output*' (economic impact); the best alternative selection is $X_3 - X_{1.}$ (by a substantial difference) (Table 5.5),
- From the point of view of objective function Zb, that is '*Minimization of air contamination*' (environmental impact), best selection is $X_3 - X_1$ (by a substantial difference) (Table 5.5),
- From the point of view of objective function Zc, that is '*Maximization of rural electrification*' (economic and social impact), the best option is $X_1 - X_3$ (by a substantial difference) (Table 5.5),
- From the point of view of objective function Zd, that is '*Maximization of job generation*', the best selection is $X_3 - X_1$ (by a substantial difference) (Table 5.5),
- From the point of view of objective function Ze, that is '*Minimization of purchased energy*', the best selection is $X_1 - X_3$ (by a small difference) (Table 5.5).

As a bottom line it seems logical to assert that the best selection is $X_3 - X_1$ especially considering that in the case of Ze (energy extracted from the grid or energy import), the advantage of X_1 over X_3 is reduced.

5.8 Conclusion of This Chapter

The aim of this chapter has been to demonstrate in an actual case the potential of Linear Programming and the wealth of information it can provide. The case consisted of three different technical options for power generation, with five objectives and subject to many criteria, however, only the five objectives (targets) have been considered because our interest was in concentrating the analysis on the different objectives proposed. As explained, the model has been solved, using in turn one target as objective function, while the other targets were considered just criteria. This procedure was repeated five times, and for each objective the following results recorded:

- Results obtained for each target when one of them in turn is considered as the objective function,
- Values of the alternatives according to each objective.

The analysis of these results has led to:

1. Determining the best blend of alternatives and its relative importance when considering all objectives proposed,
2. Supplying the DM with very detailed information regarding the impacts of each objective, which also leads to getting the ranking of the different objectives proposed,
3. Quantitative information about how each objective impacts on the others.
4. A sensitivity analysis for each objective which showed the two main targets –amongst the five proposed – with more influence,
5. Allowing the DM to realize that the popular belief and political posturing about restricting energy importation was wrong, which called for position change. This is very important because it was demonstrated that a decrease in importing energy would damage the local economy instead of benefitting it; this is a conclusion that cannot be reached unless an analysis of this type is conducted,
6. Responding to a series of logical and reasonable questions posed by the DM.

As a summary this chapter showed how a complex problem can be solved by LP and especially how to manage the information it provides.

References

*Chankong, V., & Haimes, Y. (1983). *Multiobjective decision making – Theory and methodology*. New York: Elsevier Science.

*Forman, H., & Selly, M. (2001). *Decision by objectives. How to convince others that you are right*. Singapore: World Scientific Publishing.

*Indicates suggested reading material not mentioned in text.

* French, S. (1988). *Decision theory: An introduction to the mathematics of rationality*. Chichester: Ellis Horwood.
* Hobbs, B., & Horn, G. (1997). Building public confidence in energy planning: A multimethod MCDM approach to demand-side planning at BC Gas. *Energy Policy, 25*(3), 357–375.
* Korhonen, P., Moskowitz, H., & Wallenius, J. (1992). Multiple criteria decision support – A review. *European Journal of Operational Research, 63*, 361–375.
Kumar, S., Hanna, A., Natarayan, P. (2003). Application of fuzzy linear programming in construction projects. *International Journal of IT in Architecture, Engineering and Construction (IT-AEC)*, *4*(1), 265–274. Chimay Anumba (Ed.)
* Luce, R., & Raffia, H. (1957). *Games and decisions*. New York: Wiley.
* Maino, M., Pittet, D., Kobrich, G., Claus, G. (1993). *Un instrumento para el diseño de sistemas de producción*. Santiago de Chile: Centro Latinoamericano para el Desarrollo Rural (RIMISP).
* Mészáros, C., & Rapcsák, T. (1996). On sensitivity analysis for a class of decision systems. *Decision Support Systems, 16*, 231–240.
* Ríos, S. (1980). *Decisiones multicriterio con ordenaciones parciales*. Tesis doctoral, Editorial de la Universidad Complutense de Madrid.
* Romero, C. (1981). El enfoque multiobjetivo en los modelos matemáticos de planificación de cultivos. *Revista de Economía Política, 789*, 179–204.
* Romero, C. (1996). *Análisis de las decisiones multicriterio*. Madrid: Isdefe.

Chapter 6
The SIMUS Method

Abstract The SIMUS method is not too different from what has been analyzed so far. However, it plays a part by providing a means to work with multiple objectives – no matter how many – in order to give, if not an optimal solution as in the cases that are described in Chaps. 4 and 5, a satisfactory result which considers all the objectives proposed, and produces a ranking of the different alternatives involved. SIMUS also develops a methodology to work with subjective criteria, something that is not allowed in the conventional LP method. SIMUS is demonstrated with an actual example which is complex enough to appreciate what the model can offer.

Keywords Multiobjective structure • SIMUS • Thresholds • Objectives and attributes • Normalization

6.1 Linear Programming and Its Use for Solving Multicriteria Problems

LP was the first method used for decision-making since its conception by Leonid Kantorovich (1939), and for which he was awarded the Nobel Prize in Economy in 1975 along with Tjalling Koopmans, for their development of the *"Theory of optimal allocation of resources"*. Note that in essence, this is the very nature of our problem that is to select projects or their alternatives which are subject to limited resources such as funds, time, manpower, transportation, etc., as well as the compliance of restrictions imposed by the environment, social conditions, and sustainable issues, amongst others. However, what we want in general is that the selection of projects or alternatives be made not aiming at an optimal output, but to a satisfactory allocation of our resources and compliance with restrictions.

Kantorovich ideas were further expanded by George Dantzig, who developed the Simplex algorithm to solve linear problems subject to linear restrictions and with a

N. Munier, *A Strategy for Using Multicriteria Analysis in Decision-Making: A Guide for Simple and Complex Environmental Projects*, DOI 10.1007/978-94-007-1512-7_6,

single objective function. Later on, Charnes et al. (1961) and others tried to apply the method to solve problems with various objectives which led to the development of several algorithms such as *'GOAL PROGRAMMING'*, which can treat this kind of problems. This inability of LP to treat multiobjective problems led to the development of Multicriteria Decision-Making methods, known as MCDM. The SIMUS method, described in this book is an attempt to deal with these difficulties inherent to LP, and make it suitable to solve problems with uncertainties and with multiple objectives.

6.1.1 Conditions to Meet in Decision Problems

In general, in decision problems where there are several alternatives subject to restrictions, criteria or targets, there are six components (Keeney and Raiffa 1976; Pitz and y McKillip 1984). They are:

1. *An objective or group of objectives or purposes, ends,* etc., *that one wishes to achieve,*
2. *The group of DM entity who establish their preferences regarding which will be the evaluation criteria to be used,*
3. *A set of evaluation criteria (objectives or physical attributes),*
4. *A set of alternatives, programs or options,*
5. *A set of non-controlled variables such as nature's states,*
6. *A set of results or associated values (scores) with each one of the alternatives.*

The LP model complies with them all.

According to Martel (1987), from Moez et al. (1998), *"The application of a multicriteria approach usually requires the following steps:*

1. *To define the list of actions or potential solutions,*
2. *To define the list of relevant criteria,*
3. *To evaluate the potential behaviour of each action for each relevant criterion,*
4. *The aggregation of these individual components and determination of the action that best satisfies the DM entity".*

Martel also clarifies that these criteria refer to enterprises and services with portfolio administration services, and this is a subject where LP is recognized as having a very large field of action.

6.1.2 LP as a Suitable Tool to Solve Multiobjective Problems

It is argued that LP, with its unique objective, is not a suitable tool for multiobjective analysis, nonetheless, the examples solved in this book proves that this concept is not entirely correct. It is a tool that can be adapted not for finding the optimal solution for a multiobjective problem (which otherwise is something that many DMs do not care about), but by using a methodology supporting the decision-making entity in the analysis of a problem in comparing different satisfactory solutions (Matsuhashi 1997; Ignizio 1994). This characteristic of LP is able to give the DM the possibility of formulating a series of remarks and reasoning such as: *"Fine, if what we want is to get the maximum benefit, then we must select project D. If we want to reduce to a minimum the environmental damage, we should pursue project F (or perhaps the same D). However, if we want to maximize the number of persons benefited with this project, then we must choose A,* etc."

Consequently, it is possible to deliver a portfolio of alternatives, complying with all restrictions of the problem, whatever they might be, for the DM to select the one according to his/her priorities. Naturally, it can be argued that in this case not all the objectives are considered simultaneously, which is true, but it is also worth remembering that it is not possible to get a Pareto optimum when this condition of simultaneity is established, whatever the method used.

6.1.3 Criteria and Thresholds

Criteria in LP must have limits (thresholds), and this is one of the arguments put forward by opponents of using LP for decision-making, since it is argued that thresholds limit the choices of the DM, who normally wishes to impose his/her points of view. There is no doubts that the DM labor is fundamental and the most important task in the decision-making process, but it is also necessary to remember that he/she must coexist with reality, where everything is limited, such as land, water or funds, and not rely only on good wishes which can be more or less reachable. From this point of view thresholds are fundamental, since it is not enough to say for instance that an objective calls for the whole population to have abundant and clean water, because it is necessary to establish lower and upper limits in order not to deprive the citizens with this invaluable and scarce asset, but also to avoid its squandering.

Indeed, in LP criteria must be restricted in their scope, but apart from being a fact of life, it is also done in the other methods albeit in a different way. Assume for instance that several options are analyzed on the basis of a certain criterion, say for instance *'Funds availability for a project'* which of course means establishing a value for said project. If two or more alternatives offer estimates that are above this established budget, it is obvious that these options are not to be considered and must be discarded, because they do not comply with said criterion, and this is equivalent to the LP restriction when it has a '≤' operator.

By the same token, if options are analyzed on a criterion that establishes a production, whatever it might be, to be above a certain minimum value, the DM will discard alternatives that do not comply with this requirement. This is equivalent to the LP restriction when a criterion has an operator type '≥'.

In general conventional methods do not use thresholds, but as we can see, in this respect, there are more similarities than differences between LP and that type of methodologies, although procedures are totally distinct.

6.1.4 Objectives, Criteria and Solutions

Although it is possible to change objectives into criteria (or targets) and these into objectives, there is a basic difference between the *'objective'* and *'criteria'* concepts. In general, objectives express attainable goals but are imprecise in their values, such as for instance *'Minimize costs'* or *'Maximize number of benefited people'*, or *'Minimize environmental impact by visitors in a cave'*. However, these objectives transformed into criteria would be *'Cost must not exceed 240,000 Euros'*; *'Number of benefited people must be a minimum 450'* or *'The number of visitors to the cave should be less than 50 persons per day, to preserve humidity'*, and this is the reason for using thresholds.

It is possible besides, that various objectives derive from the same main objective function. Replicating the example proposed by Cohon (2003), for an objective function such as *'Maximization of social welfare'*, there can be three different and conflicting sub-objectives such as *'Economic efficiency'*, *'Environmental quality'* and *'Equity'*. It would rather be difficult for a participant not to agree with this objective function, but it is also necessary to recognize that there could be many different opinions, including contradictory, about the sub-objectives and their quantification, and also perhaps with its proper existence.

6.1.5 Comparing Results Between Methods

There is no method that can guarantee an optimum solution with multiple objectives, and there is no way to ascertain the best solutions against which compare results obtained. However, considering that all methods start with the same data, aim at the same final purpose, and follow sound mathematical procedures, the results should be equivalent. Consequently, if a problem is solved by different methods and their results coincide, it would indicate that the solution found is probably the best. This is confirmed in this chapter when an actual problem is solved by different methods, with coincidence in the final results. However, it is not easy to compare results to the same problem solved by different methods, because the assumptions made in some of them may be incorrect and especially due to the necessity of using subjective values. LP does not have that problem, because other than the selection of criteria – common to all methods – there are not subjectivities, as for instance in establishing weights for criteria, because the model does not need them.

Regarding the potential similarity of results amongst different methods or models when compared with LP these scenarios exist:

1. *There could be full concordance of selected alternatives and in their ranking or ordering.*
   In this circumstance, it is reasonable to think that if, even working with a unique objective function and without weights in the criteria, LP is able to reach in many cases the same results as the multiobjective methods – naturally using the same data and the same restrictions – this would seem to indicate that LP has the ability to treat as restrictions or 'targets' all objectives posed, in the same way as other methods, and getting similar results. Consequently, it can be asserted that LP possesses at least, the same capacity of the multiobjective methods to solve the problem.
2. *There is concordance in the first selection and may be also in the second.*
   In this case LP ranking differs from that obtained through other methods. But, it is necessary to remember that LP with only one objective produces optimal results (while SIMUS renders satisfactory solutions with several objectives), and these results are independent of whoever performs the computation. Unfortunately it is not possible to assert the same for other methods, because their results depend on DM's opinions, ideas and preferences, and thus they can change if another person analyzes the same problem.
3. *There is no concordance at all*
   Naturally, it can be the consequence of a faulty formulation in any of the conventional methods, and LP included, for even if a method gives a solution to the proposed mathematical model, it does not necessarily mean that the problem has been correctly set up; maybe there is a non-realistic assumption or simply a mistake. Obviously, it can also happen that the solution from the other method is incorrect for whatever reasons.

6.2 The SIMUS Method – Procedure to Solve Multiobjective Problems

When a multiobjective problem is set up to be solved by SIMUS, the procedure is the same as explained in Sect. 4.4 and Chap. 5, and all the analysis done there is also applicable here. The decision matrix is the same as before with all criteria pertaining to the project, but now the proposed objectives are added as targets, and as a matter of fact, all criteria can be used as targets and then as objective functions, if wished.

As an example to illustrate SIMUS consider a problem with three alternatives (A, B, C), three objectives, and ten criteria. To solve it proceed as follows:

1. Build the decision matrix with the three objectives added as targets (Fig. 6.1).
2. Select a target to be an objective, such as target 1. This is extracted from the decision matrix and used as objective function 1, as shown.

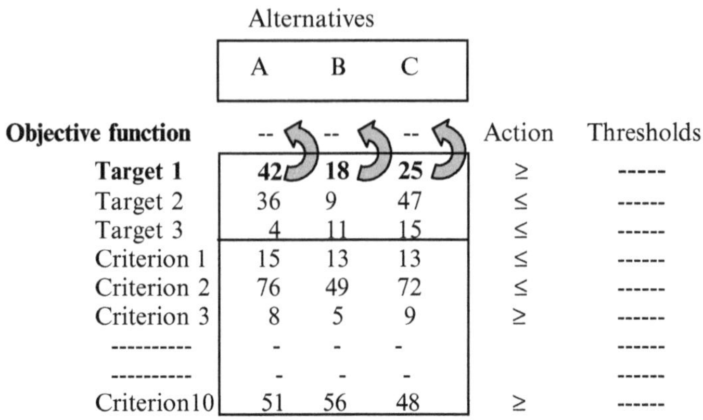

Fig. 6.1 Extraction – Target 1 is extracted to perform as objective function

3. Run the Solver and place results or scores for alternatives in the payoff matrix (Table 6.1). However, using an objective function does not guarantee that all the alternatives will get a value, because sometimes an alternative is left out the solution, as seen in Chap. 5. Consequently, in this example there could be a row with 3, 2, 1 or 0 values. The last case corresponds when there is no mathematical solution using a certain objective function.

Table. 6.1 Structure of the payoff matrix

	Alternatives		
	A	B	C
Target 1	----	-----	-----
Target 2			
Target 3			

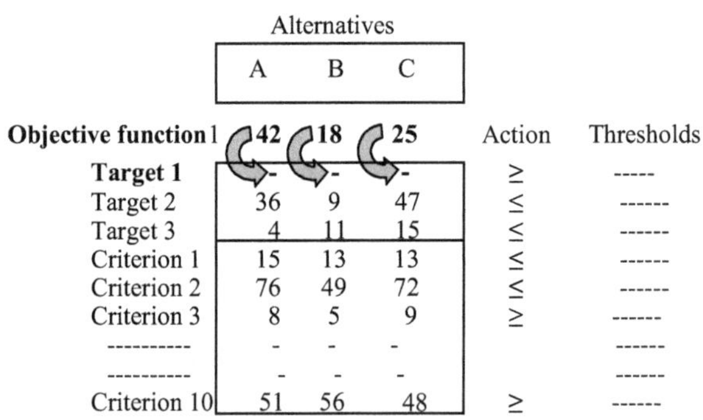

Fig. 6.2 Restoration – Objection function 1 is restored as target 1

4. Restore the just used objective in the decision matrix (Fig. 6.2)
5. Choose another target, for instance target 2 (Fig. 6.3). Repeat the procedure and place the result in the payoff matrix.
6. Restore the just used objective, select a new target, and repeat the procedure until all intended targets have been used as objectives functions.
7. Apply guidelines 1 and 2 to the payoff matrix.

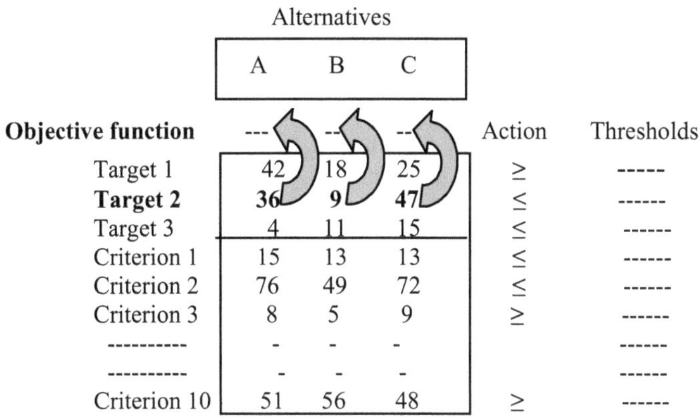

Fig. 6.3 Extraction – Target 2 is extracted to perform as objective function

Guideline 1:

It aims at determining the *importance* or worth of the alternatives, by adding up the scores found for each one. Proceed as follows:

- Normalize scores along each row of the payoff matrix.
- When complete, add up scores in each column, i.e., for a variable or alternative 'j' and 'n' targets, find this expression:

$$\text{SUM} = \sum_{i=1}^{n} Xij$$

Guideline 2:

It aims at considering the level of *participation* of each alternative regarding number of targets.

In the above scheme there are three targets, and values have been found for alternatives corresponding to each target, when it is used as an objective function. It could be that an alternative gets a higher SUM – outranking other alternatives – merely because it has a large score in one or two targets, while other alternatives get lower SUM values albeit they participate in more targets but with smaller values. It is obvious that an alternative that has values in many targets is more valuable than another that has values in fewer, since it demonstrates that said alternative is chosen in various different scenarios.

For this reason it is now necessary to compute a *'Participation ratio'* (PR), which simply is the ratio between the number of targets where an alternative has values and the total number of targets.

Therefore the next step is to compute these PR for each alternative and then normalize them (NPR).

Last step is to multiply (SUM) and (NPR), and this product will show the final importance of each alternative. Alternatives are selected according to their values, the higher the better, and then it is possible to have them ranked.

Guideline 3 is optional, and it is just another way of determining ordering or ranking. It does not relate to the ones just commented on and it uses a different way of appraisal. It consists in determining the dominance of each alternative over the others, through an analysis of the payoff matrix, by examining each normalized row and identifying the largest value, which in turn identifies the dominant alternative for that objective.

When this analysis is complete for each row, it is possible to build a *'dominance matrix'*. This is a square matrix with both columns and rows containing alternatives and where each row determines dominance over columns. That is, if alternative A has a larger value than alternative B, place a '1' in row A and column B. Repeat the processes for all rows and when completed, add up the values in rows and in columns. The resulting figure will indicate the number of times that an alternative in a row is larger than other alternatives in columns. By the same token, the resulting values for each column indicate the number of times that an alternative in a column is outranked by other alternatives in rows.

In this way two vectors are obtained, a column vector at the right that identifies *dominant* alternatives, and a row vector at the bottom, that identifies *dominated* alternatives. Then, deducting from the column vector corresponding to an alternative the value corresponding to the same alternative from the row vector, the net difference indicates its relative worth, and from here the ranking.

It is understood that there is no reason for strict correspondence between the results achieved with guidelines 1 and 2 and those from guideline 3, since guideline 1 works with score *sums*. Guideline 2 deals with the *ratio* between the number of times an alternative participates in targets, and total number of targets, and affects the results from guideline 1, and Guideline 3 considers *dominance* relationships between alternatives. Nevertheless, an analysis over 25 projects carried out by this author, using both procedures, gave complete agreement in 20 cases, that is 80%.

6.3 Case Study: Airport Expansion Plan (Solved with SIMUS)

This actual case study was proposed, set up and solved by its authors using a model called 'Regime'.

Case: ***'A Multicriteria Decision Support Methodology for Evaluating Airport Expansion Plans'***
Authors:

Ron Vreeker
Peter Nijkamp
Chris Ter Welle

Organization: **Tinbergen Institute**[1] **– Discussion Paper TI 2001-005/3**

[1]The Tinbergen Institute is the Institute for Economic Research of the Erasmus Universiteit Rotterdam, Universiteit van Amsterdam and Vrije Universiteit Amsterdam.

Peter Nijkamp kindly granted authorization to utilize this example.

Brief description

The example deals with the potential expansion of an airport in the Maastricht area, The Netherlands, for which four alternatives are considered, and identified as:

Alternative A: Status quo (Do nothing – Business as usual),
Alternative B: The airport will concentrate as a passenger airport,
Alternative C: The airport will be a regional airport,
Alternative D: Permits commercialization. The airport can purchase carbon credits[2] from other airports in order to be able to increase the volume of its commercial operations.

There are 20 criteria to evaluate alternatives, all of them qualitative and calling for maximization. Criteria have been chosen in accordance with the sustainability principle and thus they consider the economic aspect (criteria 1–8), as well as the social (criteria 9–14), and the environment (criteria 15–20). It is a well-structured real life example, complex enough, and it will be used to illustrate SIMUS's applicability.

6.3.1 Problem Set-up and Construction of the Decision Matrix

Table 6.2, the decision matrix, shows the original values, which are the contributions of each alternative to each criterion. Thus, the contribution of alternative B to criterion 3 for instance, is 9. The last column at the right includes the summation of all the values in the corresponding row, which will be used for normalization purposes.

6.3.2 Normalization

In order to work with homogenous quantities, values in Table 6.2 are normalized (this is mandatory in SIMUS), by dividing each one by the summation of values for each row, and thus Table 6.3 is built. Once this normalization is complete, compute the independent terms vector using for each row the maximum value of said row. Then, in row 5 for instance, the independent term is 0.400, which is the maximum value of that row. However, this is because we are maximizing; in a minimization case, the chosen value for the independent term will be the lower of the row. SIMUS, as others Linear Programming models, can work with any combination of maximization or minimization criteria.

---

[2]Carbon credits. Some governments establish limits for contamination particularly for CO_2 emissions from industrial companies or undertakings generating emissions. If a company needs to produce more goods or provide more services but is constrained by these limits, it can offset its emissions by purchasing permits from other companies and using them. This can be done through dedicated carbon exchange markets in the USA and in Europe.

The ranking reached by the authors is:

B – C – A – D

Table 6.2 Decision matrix

Criteria		Alternatives					
ID	Description	A	B	C	D	Objective's action	Sum of scores in each row
1	Economic benefits to region	4	8	9	6	Max	27
2	Employment in transport and logistic sectors	4	5	8	6	Max	23
3	Employment in financing and business sectors	5	9	7	6	Max	27
4	Employment in tourism and recreation sectors	6	5	1	3	Max	15
5	Development and logistics of industrial sites	5	3	1	6	Max	15
6	Infrastructure	5	8	9	6	Max	28
7	Business traffic	6	9	8	6	Max	29
8	Supply of skilled jobs	4	7	9	6	Max	26
9	Noise	5	8	5	3	Max	21
10	Safety	5	7	2	3	Max	17
11	Health	5	6	5	3	Max	19
12	Recreational traffic	6	9	9	7	Max	31
13	Total income	4	6	7	6	Max	23
14	Residential areas	5	5	2	5	Max	17
15	Natural conservation areas	5	4	1	3	Max	13
16	Air quality	4	6	1	4	Max	15
17	Water quality	4	5	2	4	Max	15
18	Soil quality	4	3	2	4	Max	13
19	Biodiversity	4	5	2	4	Max	15
20	Habitat disturbance	5	8	1	3	Max	17

Source: Adapted from authors' *Table 3. 'The impact matrix for alternative airport expansion plans'*

6.3.3 *Model Operation*

All of these criteria can be used as objective functions; in this case we choose 12 criteria or targets[3] out of 20, and utilize them consecutively to perform as objective functions. We will have then 12 targets, and these are I.D. 1-2-3-4-9-20-12-13-19-10-11-5. Observe that it is irrelevant the order in which targets are selected and run. If there were more objective functions they can be added to the decision matrix, as exemplified in Fig. 6.1.

[3] The words 'criteria' ore 'target' are equally used and with the same meaning.

Table 6.3 Normalized values of Table 6.2

Criteria/targets		Alternatives						
ID		A	B	C	D	Action	Op.	Thresholds
1	Economic benefits to region	**0.48**	**0.96**	**0.33**	**0.222**	Max	≤	**0.333**
2	Employment in transport and logistic sectors	**0.174**	**0.217**	**0.348**	**0.261**	Max	≤	**0.348**
3	Employment in financing and business sectors	**0.185**	**0.333**	**0.259**	**0.222**	Max	≤	**0.333**
4	Employment in tourism and recreation sectors	**0.400**	**0.333**	**0.067**	**0.200**	Max	≤	**0.400**
5	Development and logistics of industrial sites	**0.333**	**0.200**	**0.067**	**0.400**	Max	≤	**0.400**
6	Infrastructure	**0.179**	**0.286**	**0.321**	**0.214**	Max	≤	**0.321**
7	Business traffic	**0.207**	**0.310**	**0.276**	**0.207**	Max	≤	**0.310**
8	Availability of jobs demanding abilities	**0.150**	**0.269**	**0.346**	**0.231**	Max	≤	**0.346**
9	Noise	**0.238**	**0.381**	**0.238**	**0.143**	Max	≤	**0.381**
10	Safety	**0.294**	**0.412**	**0.118**	**0.176**	Max	≤	**0.412**
11	Health	**0.263**	**0.316**	**0.263**	**0.158**	Max	≤	**0.316**
12	Recreational traffic	**0.194**	**0.290**	**0.290**	**0.225**	Max	≤	**0.290**
13	Total income	**0.174**	**0.261**	**0.304**	**0.261**	Max	≤	**0.304**
14	Residential areas	**0.294**	**0.294**	**0.118**	**0.294**	Max	≤	**0.294**
15	Natural conservation areas	**0.385**	**0.308**	**0.077**	**0.231**	Max	≤	**0.385**
16	Air quality	**0.267**	**0.400**	**0.067**	**0.267**	Max	≤	**0.400**
17	Water quality	**0.267**	**0.333**	**0.133**	**0.267**	Max	≤	**0.333**
18	Soil quality	**0.308**	**0.231**	**0.154**	**0.308**	Max	≤	**0.308**
19	Biodiversity	**0.267**	**0.333**	**0.133**	**0.267**	Max	≤	**0.333**
20	Habitat disturbance	**0.294**	**0.471**	**0.059**	**0.176**	Max	≤	**0.471**

The first objective will be target I.D. number 1, '*Economic benefits to region'*, which is removed from the decision matrix and used as objective function, and calling for maximization. Solver is run, and the result (i.e. the scores for each alternative), placed in the first row of the payoff matrix (Table 6.4), which corresponds to this objective function (In this case, Solver found only one value for alternative C). Once this value is logged, criterion '*Economic benefits to region*' is restored into the decision matrix.

The second objective function corresponds to target '*Employment in transport and logistic sectors*', which is removed from the decision matrix and inputted in Solver as objective function. Solver is run, and a new set of values obtained, which are placed in the second row of Table 6.4. Again, Solver selected only one alternative. Once this value is logged, criterion '*Employment in transport and logistic sectors*' is restored into the decision matrix.

The third objective function corresponds to criterion '*Employment in financing and business sector*', which is removed from the decision matrix and inputted in Solver as objective function to be maximized. Solver is run and a new result obtained, which is placed in the third row of Table 6.4. Once this value is

Table 6.4 Payoff matrix

Selected objectives		Alternatives A	B	C	D
1	Economic benefits to region			1.00	
2	Employment in transport and logistic sectors			1.00	
3	Employment in financing and business sectors		1.00		
4	Employment in tourism and recreation sectors	0.608	0.429		0.071
9	Noise	0.162	0.192	0.063	0.113
20	Disturbance of fauna habitat			1.00	
12	Recreational traffic	0.233	0.325	0.557	
13	Total income		1.00		
19	Biodiversity	0.393	0.510	0.086	0.181
10	Safety		1.00		
11	Health	0.564	0.498	0.125	
5	Development and logistics of industrial sites				1.00

logged, criterion *'Employment in financing and business sector'* is reinserted into the decision matrix.

The procedure is repeated until all selected criteria or targets are used as objective functions and further restored into the decision matrix. Observe that for some criteria Solver shows values for the four alternatives, and in other for two or three. The absence of values for a certain objective and a particular alternative, as for instance alternative C in *'Employment in tourism and recreation sectors'*, means that said alternative is not included in the set of optimal solutions for that objective function. That is, alternative C is not part of the solution, implying that it is not relevant for that objective, even if the alternative has a score for that target.

There is little chance that in an actual case like this, all 20 criteria may be considered as objectives, or even 12, as is shown here, but it was done just to illustrate that there is no limit for the number of objectives taken into account.

Once completed the payoff matrix it is normalized and results shown in Table 6.5. Guideline 1 is now applied and its results shown in the SUM row. Since alternative B has the largest value (4.887), it is the best alternative according to this guideline; alternative C with a score of 3.782 is the second, and so on.

Now we use Guideline 2 to compute the participation ratio, which is depicted in (PR) row. Notice that alternative A for instance, intervenes or participates in 5 out of a total of 12 targets, therefore its PR = 5/12 = 0.146. Row (NPR) shows these ratios normalized, by dividing each one by their sum, which is 2.081. Then (NPR) for alternative D for instance will be 0.416/2.081 = 0.199.

Row (SUM) x (NPR) is the solution, and it is seen that the 'best' or first alternative is B, the second C and so on.

Ranking is then:

B – C – A – D

This ordering is the same as found by the authors using the Regime model; however, there is no guarantee that the result from one method coincides with results

Table 6.5 Normalized scores for alternatives obtained from different objective functions

Selected objectives		Alternatives			
		A	B	C	D
1	Economic benefits to region			1.00	
2	Employment in transport and logistic sectors			1.00	
3	Employment in financing and business sectors		1.00		
4	Employment in tourism and recreation sectors	0.549	0.387		0.064
9	Noise	0.306	0.362	0.119	0.213
20	Disturbance of fauna habitat			1.00	
12	Recreational traffic	0.202	0.282	0.484	0.031
13	Total income		1.00		
19	Biodiversity	0.336	0.436	0.074	0.155
10	Safety		1.00		
11	Health	0.475	0.420	0.105	
5	Development and logistics of industrial sites				1.00
Guideline 1 (SUM)		1.868	4.887	3.782	1.463
Guideline 2 (PR)		5/12 = 0.416	8/12 = 0.666	7/12 = 0.583	5/12 = 0.416
(NPR)		0.199	0.320	0.280	0.199
(SUM) × (NPR)		0.371	1.564	1.058	0.291
Alternatives ranking		Third	First	Second	Fourth

from another, because they are based on different principles and assumptions. Nevertheless, the fact that in this case there is an absolute coincidence between both methods when there are 4! = 24 possible combinations, seems to indicate that this result might be the best.

6.3.4 Using the Third Guideline

Use Table 6.4 for reference of scores of alternatives.

Objective 1:
Clear dominance of alternative **C** over all others. Then, a '1' is placed in cells **C**-A, **C**-B, and **C**-D on *'Dominance matrix'*, Table 6.6.

Objective 2:
Clear dominance of alternative **C**. Repeat the same procedure in Table 6.6 adding a new '1' in cells **C**-A, **C**-B, and **C**-D.

Table 6.6 Dominance matrix

	A	B	C	D	Total by row	Total by row MINUS total by column	Net dominance (difference between row and column values for the same I.D.)
	1	**2**	**3**	**4**	**5**	**6**	**7**
1 **A**		1+1	1+1	1+1	(6)	**6 − 10 = − 4**	**Third**
2 **B**	1+1+1 +1+1		1+1+1 +1+1	1+1+1 +1+1	**15**	**15 − 7 = 8**	**First**
3 **C**	1+1 +1+1	1+1 +1+1		1+1 +1+1	**12**	**12 − 8 = 4**	**Second**
4 **D**	1	1	1		**3**	**3 − 11 = − 8**	**Fourth**
5 **Total by column**	(10)	7	8	11			

Objective 3:
Clear dominance of alternative **B**. Place '1 s' in cells **B**-A, **B**-C and **B**-D in Table 6.6.

Objective 4:
Dominance of alternative **A**. Place '1 s' in cells **A**-B, **A**-C and **A**-D in Table 6.6.

Objective 9:
Dominance of alternative **B**. Place '1 s' in cells **B**-A, **B**-C and **B**-D of Table 6.6.

Objective 20:
Clear dominance of alternative **C**. Place '1 s' in cells **C**-A, **C**-B and **C**-D of Table 6.6.

Objective 12:
Dominance of alternative **C**. Place '1 s' in cells **C**-A, **C**-B and **C**-D of Table 6.6.

Objective 13:
Clear dominance of alternative **B.** Place '1 s' in cells **B**-A, **B**-C and **B**-D of Table 6.6.

Objective 19:
Dominance of alternative **B**. Place '1 s' in cells **B**-A. **B**-C and **B**-D of Table 6.6.

Objective 10:
Clear dominance of alternative **B**. Place '1 s' in cells **B**-A. **B**-C and **B**-D of Table 6.6.

Objective 11:
Dominance of alternative **A**. Place '1 s' in cells **A**-B, **A**-C, and **A**-D of Table 6.6.

Objective 5:
Clear dominance of alternative **D**. Place '1 s' in cells **D**-A, **D**-B and **D**-C of Table 6.6.
Add up values in each row and in each column as shown in column 5 and row 5.

Find the differences between rows and columns for the same alternative. Thus for objective B for instance it will be 15 − 7 = 8.

The Net Dominance vector shows that objective B has the largest value.
Then, the ranking is:

B – C – A – D

It can be seen that there is a complete agreement with the ranking found using guidelines 1 and 2, and with the original work.

6.4 Conclusion on This Example

The actual and relative complex example proposed demonstrates how LP is an appropriate decision-making tool when there are multiple and even contradictory objectives, analyzing ceteris paribus a certain objective function. Again, LP cannot solve a problem with many objectives although some methodologies such as GOAL PROGRAMMING (Spronk 1981), the *'Restrictions Method'* and others can work out these type of problems, but with few objectives. The purpose of this work is to show a new approach that, based on LP, can determine a set of values to reach solutions that the DM can use.

The main advantage of this method is that it delivers a reduced set of optimal solutions, as many as the number of objectives established, and gives the necessary numerical values for a qualitative and quantitative analysis, as has been exemplified. Again, it does not provide an optimal global solution, but instead a wealth of reliable and objective information, and giving the DM the chance to examine several scenarios of the type *"What happens if......?"*.

Furthermore, notice that the method does not use subjective values, other than an assumption for selecting criteria – common to all methods – and thus ensuring a more reliable answer. On the other hand, it is worth remembering that the final set of solutions derives from optimal values of the alternatives, which adds more assurance that results are possibly more reliable than in other methods.

6.5 The Dual in Linear Programming and Its Application to Projects

Any LP problem is a *'primal'* and its 'mirror image' a *'dual'*. Both are linked as per the relations depicted in Table 6.7.

Duality in LP is more than a mathematical curiosity, since in reality it is a mathematical phenomenon which simulates different economic problems and hence its importance Suarez (1980). It is vital to mention that the primal and its dual offers the same value for the objective function. This is not a mere coincidence but a perfect correlation founded on the Duality Fundamental Theorem.

Table 6.7 Equivalencies between primal and dual

Primal		Dual
To a restriction	Corresponds	A variable
To a variable	Corresponds	A restriction
To an independent term	Corresponds	An objective coefficient
To an objective coefficient	Corresponds	An independent term
To an inequality	Corresponds	An inequality
To an equation =	Corresponds	An inequality
To an objective maximization	Corresponds	An objective minimization
If $x_j > 0$	Corresponds	An equation
To an inequality ≤	Corresponds	A variable equal to 0

6.6 Sensitivity Analysis

All multicriteria methods supply a base for decision-making; LP is no exception, and even if its result is a mathematical optimum it does not necessarily mean that the found value is the most convenient or amenable. It simply constitutes a reference framework to allow the DM to analyze a selection that says which is the best solution from the mathematical point of view, which may not coincide with the operative optimum, but it provides a yardstick against which to compare conclusions.

Whatever the model utilized it is always convenient to perform a sensitivity analysis to determine how the result found will react to variations of some parameters, especially criteria and thresholds, under circumstances that originally were not considered, either for the mathematical impossibility in expressing them or because they are rooted in the DM's own experience or based on similar cases.

In most methods, this sensitivity analysis is performed altering criteria weights; however, changes in these weights, which are percentages, are relative, and consequently since their sum must equal unity, the increment or decrement of one of them forcefully implies modifying the weights of the others. Here, there are two problems:

(a) It is assumed that criteria weights are percentages and related to each other, when in reality there is no reason for that. That is, if in a project there are say seven criteria, it is possible to assign to each criterion a weight based for instance on a 1–10 scale, according to the DM's opinion and not related between them.
(b) In the percentage system a question is how to determine which criterion or criteria decrease or increment when the percentage in another criterion has been changed. It is done, but it is arbitrary.

These two problems are avoided in LP since not only does the model not need criteria weights but besides, criteria relative importance is information automatically generated by the model itself through its shadow prices of criteria.

6.7 Choosing the Final Solution

It falls now to the DM to adopt the final solution, if no more information is needed; otherwise that information must be demanded. Thus, the solution found by SIMUS, which may not be the best, is without a doubt documented and allows the DM to ask for any complementary data needed.

6.8 Conclusion of This Chapter

The SIMUS procedure and its results have been discussed in detail, demonstrating, showing that LP, complemented with an analysis of optimal solutions, allows advantageously the resolution of multicriteria problems, giving at the same time abundant information for the DM to adopt the right decision.

References

Charnes, A., & Cooper, W. (1961). Management models and industrial applications of linear programming. Wiley, New York.

Cohon, J. (2003). *Multiobjective programming and planning*. New York: Dover Publications.

Ignizio, J., & Cavalier, M. (1994). *Linear programming*. Englewood Cliffs: Prentice Hall International Series in Industrial and Systems Engineering Prentice Hall.

Kantorovich, L. (1939). *The best uses of economic resources*. Cambridge: Harvard University Press.

Keeney, R., & Raiffa, H. (1976). *Decisions with multiple objectives: Preferences and value tradeoffs*. New York: Wiley.

Martel, J-M. (1987). *Aide multicritère à la décision*. Work paper from Laval University. Website: http://www2.ulaval.ca/. Retrieved 05 Dec 2009.

Matsuhashi, K. (1997). *Application of multi-criteria analysis to urban land-use planning*. Interim Report. IR-97-091/December – International Institute for Applied Systems Analysis A-2361 Laxenburg.

Moez, H., & Martel, J.-M. (1998). *A multicriteria approach for selecting a portfolio manager*. Website: http://www.highbeam.com/doc/1P3-34927708.html. Retrieved 07 June 2010.

Pitz, G., & McKillip, J. (1984). *Decision analysis for program evaluators*. Newbury Park: Sage.

Spronk, J. (1981). *Interactive multiple goal programming for capital budgeting and financial planning*. Boston: Martinus Nijhoff Publishing.

Suarez Suarez, A. (1980). *Decisiones óptimas de inversión y financiación en la empresa*. Madrid: Ediciones Pirámide S.A.

Chapter 7
Comparison of Different Models

Abstract Many methodologies have been proposed to help the decision-making process, as has been mentioned and illustrated in Chap. 3, where four different models were examined and examples and cases posed and solved, with indication of the pros and cons of each. However, no comparisons have been made between them, and this is the purpose of this chapter. Here the first table condenses information about technical characteristics of each methodology. A second one details main characteristics from the user point of view, and a third table briefly illustrates the successive steps to use each methodology. The chapter also enters in the domain of complex projects, considering size, interrelationships between alternatives and between criteria, thresholds, etc. Finally it goes through into commenting on the advantages of Linear Programming (LP) for solving complex situations. It is understood that there is no one methodology that is better than all others, however it is also believed that LP is probably the best suited to solve complex problems.

Keywords Comparison • Complexity • Classification • Integration

7.1 Operative Differences Between Linear Programming and Other Multicriteria Methods

Linear Programming (LP) seeks to find the solution that optimizes the objective function, either maximizing it as in the case of profits or benefits, or minimizing it as in costs or damages. It is worth mentioning that LP is the only method that considers all feasible solutions; while conventional systems, such as the outranking methods or preference method, compare only pairs of alternatives and choose the dominants. Table 7.1 condenses the main technical characteristics of the diverse methodologies.

N. Munier, *A Strategy for Using Multicriteria Analysis in Decision-Making: A Guide for Simple and Complex Environmental Projects*, DOI 10.1007/978-94-007-1512-7_7,

Table 7.1 Brief on technical characteristics of diverse models

ELECTRE I, II, III, IV, IS y TRI	PROMETHEE I, II,I II y IV
It is based in a ratings scale.	It is based in a ratings scale and requires more information than ELECTRE.
ELECTRE I *selects the set of best solutions*. It works with concordance and discordance indexes. ELECTRE II aims at *doing an ordering* and establishes limits or thresholds.	The DM must decide between the different alternatives, employing not only the difference of values between them, if minimizing or maximizing, but besides must choose one of several '*transfer functions*' through some of the six types of proposed functions.
ELECTRE compares alternative A with alternative B with the purpose of accepting or rejecting and *produces a ranking*.	It talso works with preference and indifference levels.
ELECTRE IV reaches *a partial ordering*. It does not require weights for criteria. ELECTRE IS selects alternatives. ELECTRE TRI *classifies alternatives to predefined categories*. There are not interchange values between criteria. For an interesting description and comment of the ELECTRE different versions, see European Commission (2001).	There is no interchange of values between criteria.

AHP	MAUT	Linear Programming
It is based on a ratio scale. Breaks a complex problem in a hierarchical structure and works establishing multiplicative preferences scores analysing pair wise criteria. This produces a square matrix which allows the determination of the weight of each criterion using the eigenvector calculation.	It is based on computing a value function according to the following formula: $\sum_{i=1}^{m} wi.vi\left(xi(a)\right)$ Where: V(a) : Total value of project or alternative (a). v_i = Function of partial value which represents the contribution of project 'a' regarding the restriction it belongs to.	It is based on working with all restrictions, which form a polytope containing all possible solutions. Efficient solutions are in the vertexes of the polytope, and the objective function is then used to determine the optimum solution amongst all efficient solutions. This polytope is framed within the coordinate system formed by the alternatives and which can have any mathematical dimension.

Further works pair-wise comparing alternatives according to each criterion and finally applies to these results the weights computed through the eigenvalues. It is a highly regarded method because it allows working with a subjective scale and because it is easy to understand.	w_i = Restriction weight. V (a) is computed for each project and the largest is chosen, since it is assumed that if a is preferred to b, then $V(a) > V(b)$. Criteria weights should indicate to the DM the interchange values, that is how much it is gained or lost when the weight of a criterion is modified relative to others. These methods are called 'compensatory' because the low value in a criterion can be compensated by a high value in another. It can also work with probabilistic values.	The solving method is grounded in matrix algebra operations and for this reason it is difficult to understand.

As we know, either ELECTRE or PROMETHEE follow in general lines the same modus operandi, that is outranking. AHP employs preferences, and finally LP uses matrix analysis. However, there is a similitude between the outranking methods and LP since the three of them start with a decision matrix (formed by '$\mathbf{a}_{ij}$' scores) as the first step. The first two methods continue with computation of the value function '$\mathbf{f(a_j)}$' for each alternative, which is equivalent to the computation of '$\mathbf{Z}_j$' (objective function) in LP, which determines the entering variable. ELECTRE and PROMETHEE establish the dominance of one alternative over the others, while subordinate alternatives are discarded. In LP there is an equivalent action through the computation of the ratio between the independent term '$\mathbf{b}_j$' and each one of the scores '$\mathbf{a}_{ij}$' of the entering column, which is a pivot, and then determining which is the alternative that must leave the solution, and replacing it with another more efficient.

In AHP the selection of the most dominant alternative is carried out through the sum of the product of the weight of each criterion (which has been previously determined by the eigenvector method) and the weight of each alternative (also previously found by the eigenvector method), giving a valuation for each alternative, and then selecting that of the highest value.

As appreciated, ELECTRE, PROMETHEE and LP procedures are quite similar in their concepts albeit not in their mechanic, and then a logical thought leads to the notion that, when LP is applied, similar results must be expected. When this does not happen, it can be attributed to the use in outranking models of different assumptions and especially subjective estimates regarding weights and indifference and acceptance levels, and transfer functions.

This agreement in procedures is not so apparent between LP and AHP albeit it exists, although with different mechanics; consequently similar results should be obtained for the same problem. However, the fact that AHP is based on subjective preferences for the computation of weights for criteria and for alternatives – a method that is completely alien to LP – makes it difficult to find coincidences. Table 7.2 condenses the main operative differences in these systems. Table 7.3 condenses the sequential steps to develop each method.

7.1.1 Important Aspects in Project Selection

There are three factors that are important in alternatives or project selection and that not all methods consider; they are:

7.1.1.1 Complexity

Complexity in projects refers to the number of intervening elements (alternatives, criteria, thresholds, etc.), and the number of relationships between them. Probably there is more complication in relationships than in the number of elements, and indeed, when one tries to replicate as faithful as possible a real life situation, many

Table 7.2 Main working characteristics of diverse models

Characteristic	ELECTRE I, II, III, IV, IS an TRI	PROME THEE I, II,I II and IV	AHP	MAUT	Linear Programming
Comprehension difficulty	High	Medium	Medium	Low	High
Available software	Yes	Yes	Yes	Yes	Yes (several)
Effect on alternatives selection by individual objectives	No	No	No	No	Yes
Capacity, measured as number of projects that it can handle	Small number of alternatives	Small number of alternatives (Bouyssou 1990)	Small number of alternatives	Small number of alternatives	Practically unlimited
Does the method allow working with multiple objectives?	Yes	Yes	Yes	Yes	No (except in some small problems or unlimited using SIMUS)
Does the method allow establishing technical precedence of projects?	No	No	No Yes (with ANP)	No	Yes
Is it possible to pre-select a determined alternative?	No	No	No	No	Yes
Does the method allow for binary selection? (0 or 1) (to select or not an alternative)	No	No	No	No	Yes
Does the method allow for alternatives ranking?	Yes	Yes, ELECTRE II	Yes	Yes	Yes (albeit there is no guarantee because the Solver may offer a unique solution, however this is possible with SIMUS)
Does the method allow for excluding alternatives (Example: If A is chosen then B can't)?	No	No	No	No	Yes

(continued)

Table 7.2 (continued)

Characteristic	ELECTRE I, II, III, IV, IS an TRI	PROME THEE I, II,I II and IV	AHP	MAUT	Linear Programming
Number of criteria	Limited	Limited	Limited	Limited	Practically unlimited
Does the method allow for quantitative and qualitative criteria?	Yes	Yes	Yes	Yes	No with LP Yes (with SIMUS)
Capacity to limit criteria	No	No	No	No	Yes
Does the method need to establish a weight for criteria?	Yes, except ELECTRE IV	Yes	Yes	Yes	No, because LP does use them
Does the method make pair comparison of alternatives?	Yes	Yes	MAUT: Yes AHP: The method is based in pair comparison of alternatives and assumes the existence of transitivity (if C>D and D>F, then C is>F)	No	
Is there any certainty that the solution reached is the best?	No	No	No	No	Yes, but only when one single objective is used
May the method work with fuzzy logic?	Yes	Yes	Yes	Yes	Yes
Can the method work with limited resources?	Uses pseudo-criteria, but they are not thresholds to set bounds to criteria	Replaced by indifference and preference levels, but they are not thresholds to set bounds to criteria	No, unless thresholds are considered by the DM	No	Yes
Does the method allow for a sensitivity analysis?	Yes	Yes	Yes	Yes	Yes

Table 7.3 Condensed steps to utilize each model

	First step	Second step	Third step	Fourth step	Fifth step
ELECTRE	Establishes the decision table	Computes the value function of each alternative	In accordance with the value function analyses the larger. Compares pair of alternatives as a function of these values, chooses the larger and discards the balance. If the discarded value is in turn dominant of another, the second is discarded (transitive character), forming in this way the ranking or ordering.		
PROME-THEE	Establishes the decision table	Computes the value of each alternative	In accordance with values of alternatives determines the larger. Chooses two thresholds (significant (p) and indifference) (q)	Pair compares alternatives using these thresholds and according to pre-defined functions chosen by the analyst	Chooses the alternative depending the objective (maximization or minimization)
AHP	Using a dedicated numerical scale, generates for criteria a preferences square matrix that is based on the fact that if $a_{ij} = k$, then $a_{ji} = 1/k$	Computes the eigenvalue of this matrix and its result is the weighting vector for criteria	Following the same procedure pair compare alternatives for each criterion and computes and determine weights for alternatives	Weights found in second step are assigned to criteria, and then multiplies the values of alternatives' eigenvalues by these criteria weights found in third step	For each alternative sum found scores and selects those alternatives with the largest values forming the ranking in decreasing order of importance
Linear Programming	Establishes the decision table	Establishes a system of linear inequalities considering the a_{ij} values and the respective variables or alternatives. The mathematical system is expressed as $A \cdot X = B$ That is a matrix system as: This matrix arrangement is solved computing the inverse of matrix A and multiplying it by the independent term vector B.	Computes for each alternative the value of the objective function zj as the product of aij with coefficient cj, which affects the same variable in the objective function. The Simplex uses these values z_j to determine the variable or alternative entering in the new solution. For that the method finds the difference between zj and the corresponding cj, and as a function of that difference chooses the greatest or the smallest according to the kind of optimization (Max. or Min.), and introduces the corresponding variable in the new vector space. To keep the same vector space one of the alternatives must leave and it is done through the ratio between the independent terms and the respective aij of the selected column is the last step. It continues with this procedure until the optimum is reached.		

interrelationships must be considered. However, and this is the core of the problem, it is not enough to consider direct relationships but also those which are in a 'chain', and perhaps originated not by one event but for several actions operating at the same time, in sequence, in loops, or in both ways.

For instance, production of an item relates to demand for the item, which in turn may be responsive to advertising, which in turn may depend on a feedback process from low demand; but also demand depends on economic and social conditions of consumers, their willingness to pay or the amount they are willing to pay for the item, etc. On the other hand, complexity is inherent in the number of alternatives or projects competing for the same resources and trying to comply with different objectives, which can be contradictory, and for this, it is necessary to include precedence, exclusive and complementing projects, projects under execution, etc.

7.1.1.2 Case Size

Outranking and preference models have a limited scope because of the number of projects or alternatives they can treat. These methods in general give good results but in this writer's judgment, they are not suitable to manage complex situations, which precisely are those encountered in practice. In real life the analysis of a project needs to take into consideration and simultaneously – because of their interactions – social, economics, environmental and sustainability issues and conditions and restrictions involved, which are not likely to be contemplated by these methods.

7.1.1.3 Delimiting

Outranking as well as preference methods do not consider that in most cases criteria are limited by numerical values. As commented, for some researchers this is a disadvantage of LP because they believe that values turn the problem very restrictive – which is true – since effectively the problem acquires boundaries. However, criteria limits or thresholds try to replicate conditions existent in the real world, since we live in a limited environment, either referring to resources availability (labour, funds, raw materials, etc.) or natural resources (air, water, soil, etc.). There are boundaries in social issues (limits in education, in providing sanitary services, in income per capita, etc.), that is, humankind does not live in an unlimited world from any point of view. Consequently, not considering these limitations is to ignore reality.

For instance, in a problem of selecting alternative sources to generate electrical energy for a region, it is possible, using any of the techniques, to determine which alternative is preferable to others, considering quantitative factors at play, such as costs, outputs, efficiencies, etc., and other qualitative features such as people's opinion, landscape impact, social needs, etc. Then, it is possible to conclude that alternative A is better than B because it offers greater output.

However, if permissible maximum and minimum limits are not taken into account the analysis is irrelevant, because an alternative or set of alternatives can be selected, but their output could be not enough to satisfy the necessities, which are the object of the development.

If social aspects are considered this writer concurs with Munda (2004), when he says, "*During the last two decades it was understood that welfare is a multidimensional variable. This implies that a systematic evaluation of plans or projects must be based in the distinction and in the measurement of a wide set of criteria. As a consequence, the multicriteria evaluation techniques are an appropriate tool to serve as a model*", but in reality what must be analyzed is if these techniques have the capacity and ability to solve these problems, again considering the delimiting of social variables.

A survey performed by this writer shows that in general the quantity of criteria used is scarce, perhaps no more than 30, and this in some few cases. It is believed that a parallel can be made with what happened in the automobile industry, when comparing a 1950 car with last year models. At that time the main concept was a large, comfortable vehicle and aspects that nowadays are very important, like those listed below, were not even thought about.

- Components standardization,
- Emissions reduction,
- Fuel consumption,
- Noise reduction,
- Passenger safety and protections in case of impact ,
- Safety devices such as the ant blocking ABS system, etc.,
- Vehicle components recycled,
- Weight reduction and the use of lighter materials.

It is obvious then that at the present time, when purchasing a vehicle, a buyer has more elements to think about and more choices than existed 60 years earlier, and something similar happens when analyzing projects regarding land use, agriculture, product design, recycling policies, etc., which were not even considered in analyzing projects not so long ago.

Why do we comment on these examples? Because it is believed that something similar occurs in planning large projects; in the past, only a few aspects were taken into account in that activity and for that reason it was not necessary to look for sophisticated techniques for decision-making. Diverse projects were compared using, most of the time, economic criteria and thus the main concern was to select that project or alternative with the best Internal Rate of Return or Net Present Value. Later on, the social factor was incorporated and also the social evaluation of the project was included. Further on, the environmental aspect was added, and lately the sustainability axiom is mandatory in a good project selection. There is no doubt whatsoever that the complexity, size, sophistication, cost, risk, etc. of projects have increased many times over and therefore the task of selecting alternatives is much more complicated than ever.

7.2 Classification

Multicriteria decision problem is usually classified in two approaches. One of them is called MADM (Multi Attribute Decision Making), and the other is MODM (Multi Objective Decision Making). In reality, every one of these methods possesses the same structure, since all of them use multiple criteria to evaluate alternatives and forming a decision matrix. However, there are differences as follows:

MADM approach works with few alternatives, always discrete, and with a small number of criteria since their nature involves pair comparison of alternatives and then choosing the better ones, on the grounds of some rules. There is a pre-selection of competing alternatives based on technical and economic considerations, and the analyst task is to study them and find a solution that can be acceptable to the DM.

In MODM, alternatives may be continuous, as for instance in the production of two different types of paper, when the paper mill has an upper or lower total production limit in a certain period. In this case, if the upper limit is say, 600 tons per day, and when an exact weight of paper reels is not normally required, one reel can weight 30,189 kg while the next one, for another use, can weight 29,894 kg. However, alternatives in some circumstances need to be discrete as in the case for instance of selecting sites for the spatial location of a major maintenance centre for road equipment. Needless to say, in this instance it is evident that it is not feasible to build 33.78% of a plant in one site and 66.22% in another, and as a consequence, it is necessary to work with binary values, reflecting with '1', the possibility to build, and with '0' its impracticality or unfeasibility.

In general MODM applies to multiobjective problems – which is not the case of LP, which works with only one objective– but a variation of LP called '*GOAL PROGRAMMING*' does (Charnes and Cooper 1961). Ignizio (1976), also did pioneering work in industrial applications. It is interesting to point out the differences between MADM and MODM in the sense that in the first, the DM somehow explicitly chooses a pre-selected set of discrete alternatives, and works following a comparison procedure to find that or those which are dominant, that is are not outranked by any other.

In MODM, the mathematical model, through the interaction of alternatives, criteria and objective function, chooses the alternatives, i.e. they are implicit. These operating systems explain the main differences between both methodologies and how they diverge, applying MADM to smaller multicriteria, multiobjective problems which are solved through pair comparison.

MODM is usually suitable to more complex problems, with alternatives using discrete, continuous and binary values and with one or more objective functions. Differences are more formal than substantial, since in LP, as we have seen, the distinction between restrictions and objective functions is diffuse and it is difficult to determine which is which. However, the difference appears when comparing the scopes of the objective function and restrictions or targets, although both are linear functions of the same variables.

This difference is very important according to this writer's opinion; however, restrictions do not usually have fixed values, for there could also be variability in their scope (and this is shown through the inequality sign), because it is possible to use lower and upper limits. For instance, when a condition states "The amount spent must not exceed 3,563,000 Euros, because that is the ceiling of our budget, but should also be larger than 1,500,000 Euros, because that corresponds to a minimum size for a profitable operation". In this case there are limits to variability, not a diffuse aspiration. Álvarez et al. (2002), provide an excellent analysis of this subject as well as for objectives and targets.

7.2.1 Operation

Naturally, there are also weak points in LP, which is, once more, its inability to solve problems with different objectives, albeit some methods like Compromise Programming can work with a few objectives; but in reality, they have proved unusable for large and complex projects. From the point of view of available software, there are nowadays a large number of computer packages to solve LP problems and that also incorporate non-linear conditions, the latter found for instance in projects with fixed and variable costs; most allow working with integer and binary variables, which is very important. The last feature is something absent in outranking and preference methods.

The greatest advantage of LP is that it permits one to address complex problems through a mathematical model which is nearer to reality, and because of its versatility, which allows resolving a large variety of problems either simple or complex.

7.2.2 Decision-Making Strategy and Integration of Effects, Impacts and Externalities

There is another circumstance usually not considered, and it is that projects and alternatives, whatever the type they belong to, are in general not isolated units from their social, environmental and economic setting. That is, when the construction of a high speed train is studied and diverse alternative routes compared between origin A and destination B, and inclusive with other transportation forms that compete with the train, it is not possible to solely ponder the project from A and B, without considering the effect to intermediate cities C and D. Therefore, there is the need to take into account commercial, environmental and social impacts along the route.

This project will possibly have deep consequences and implications as well as ramifications in social, economic and environmental aspects. There could be different routes between A and B; a shorter route could be for the rail to go through a mountain by way of a tunnel. Another alternative might correspond to a longer route, but more economical and perhaps less damaging to the environment, for instance using a river valley, but affecting population or agricultural crops or scenery, and both must be studied contemplating all circumstances and from different angles.

Environmental problems can pop up. For example, in the alternative corresponding to tunnel construction, perhaps it entails the destruction of a valuable natural forest which is in the right of way and that constitutes the habitat of many species, or that the tracks should pass over a native cemetery, or that immediately after the tunnel there is a swamp area, which also is a tourist destination.

The other alternative, which would go through a river valley, also could present problems, like the necessity to build costly embankments for the rails running very close to the river, which overflows in summer time and that otherwise, would cover the tracks.

On the other hand, rail alternatives – and because it is a high-speed train -will leave without service a series of small cities and towns, served now by the interurban rail network, which connects them with local trains. Then it is necessary to consider the diverse externalities generated by this project. It could also provoke – and generally it happens – that the train operation will originate a tremendous decrease in air traffic between A and B, which must also be analyzed, from the commercial and social points of view, because job losses.

It is also mandatory to study the economic damage that will be generated by the laying of tracks when using cultivated soil and the compulsory monetary compensation for people relocation, which of course, increases the cost of the alternative.

Naturally, if all aspects were negative no government would build a high-speed train line. In fact, it offers many advantages, such as the large energy savings per passenger-km when compared with the number of buses needed to transport the same number of people; this is very important in countries that do not have fossil fuel sources and therefore have to import them. A project like this, as an indirect effect, can push the development of other methodologies such as the construction of wind energy farms, or photovoltaic and solar installations, which are sources of inexhaustible and clean energy, and elimination of import dependency. Other positive factors are less damage and wearing out of highways, fewer transit accidents and mainly a decrease in environmental pollution produced by buses on the road. All these benefits need also be considered together with the negative ones.

This example of the high-speed train, its advantages and disadvantages as well as direct, indirect and induced effects, is replicated in many other different types of projects, and its purpose is to make the reader aware of the facts, circumstances and aspects to consider in a decision-making strategy. All these concepts and naturally many more, are present when analyzing large projects, and all of them must be contemplated, studied and evaluated. This generates hundreds of restrictions, *but that is reality*, and the model used for decision-making has to reflect it as faithfully as possible. Figure 7.1 depicts a scheme of how a criterion can be branched to consider sub-criteria as well as limits or thresholds; in this instance, the criterion-example is about '*Contamination*'.

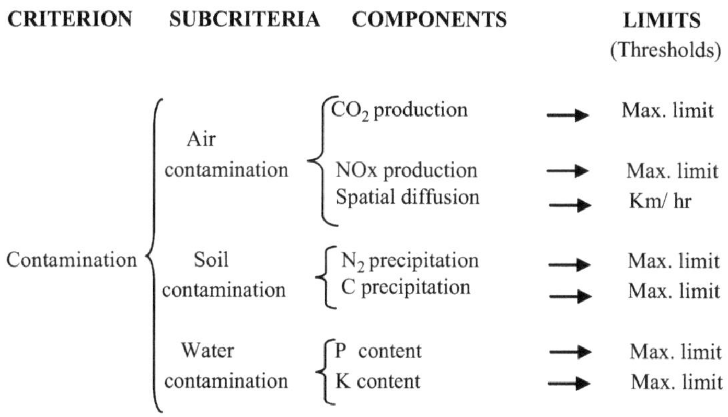

Fig. 7.1 Criterion branching, components and thresholds

7.2.3 Conclusion of This Chapter

The advantages of using LP over other methodologies for decision-making, especially on complex problems, has been documented here, showing amongst other things that LP is especially suited to treat multifaceted or intricate scenarios, many of them, in this writer's opinion, very difficult to model or unsolvable, unless LP is used. This statement is based on its ability to treat any size of problem, either in alternatives and/or in criteria – and we can be talking of thousands simultaneously – while for many of the other methods 100 or even 50 criteria or alternatives are already out of the question. LP has also the ability to establish complex relationships between alternatives such as precedence, complementarity, necessity, dependency, etc., which are unheard of in other methodologies, and last but not the least, LP's ability to work with thresholds, a fundamental issue since nothing in the world is infinite, may it be human and material resources, funds or environmental restrictions. Chapter 8, which is exclusively devoted to showing LP's efficiency in managing very complex scenarios, is the proof of these assertions.

References

Álvarez, M., Soria Alcaráz, J., García Ruiz, R. (2002). *Investigación Operativa – Modelos y técnicas de optimización*. Ed. Universidad Politécnica de Valencia.

Bouyssou, D. (1990). Building criteria: A prerequisite for MCD. In C. A. Bana e Costa (Ed.), *Reading in multiple criteria decision aids*. Berlin: Springer.

Charnes, A., & Cooper, W. (1961). *Management models and industrial applications of linear programming*. Wiley, New York.

European Commission Energy (2001). *NNE5-2001-273*: *Development and application of a multi-criteria software decision analysis tool for renewable energy sources*. (MCDA-RES).

Ignizio, J. (1976). *Goal programming and extensions*. Lexington, Lexington Books.

*Kumar, S., Hanna, A., Natarayan, P. (2003). Application of fuzzy linear programming in construction projects. *International Journal of IT in Architecture, Engineering and Construction* (*IT-AEC*), *4*(1), 265–274. Chimay Anumba (Ed.).

Munda, G. (2004). Métodos y procesos multicriterio para la evaluación social de las políticas públicas. *Revista Iberoamericana de Economía Ecológica, 1*, 31–45.

*Indicates suggested reading material not mentioned in text.

Chapter 8
Planning and Solving Actual Complex Projects

Abstract The purpose of this chapter is to show the nature of a complex project outlining the structure that many companies, entrepreneurs, government agencies, etc. have adopted. The chapter presents a series of seven actual projects executed in different countries and pertaining to different areas as diverse as river basing planning, environmental indicators, urban development, municipal policies, bridges repair scheduling, land use and metallurgical development and solved by Linear Programming. Each case has its own characteristics which are detailed and briefly commented on, so as to give the reader an idea of the potential of this tool.

Keywords Complex projects • Strategic planning • Linear programming • River basin • Environmental indicators • Housing • Municipal • Bridge scheduling • Land use • Contractors selection

8.1 The Matrix Structure of the Complex Project

Figure 8.1 shows the matrix structure of a consulting enterprise. These ventures usually execute several projects of diverse nature at the same time, but all within the realm of professional activity of the firm. Thus, a large consulting firm may be designing pulp and paper mills, metallurgical plants, harbours, or environmental rehabilitation undertakings. Generally, they work with a matrix structure having in Level 1 the Direction with its Board of Directors, and with a General Manager.

Level 2 pertains to all the firm's functions which are shared by all projects – some being global services – and involving, Human Resources, Financing, Engineering, Administration, Informatics, Design, Quality Control, etc. For a given project, say the construction of a paper mill, all functional departments of the firm support this project, consequently, Human Resources is in charge of personnel recruitment. Financing executes the economic and financial analysis, while Engineering is responsible for the technical development of the project, machine selection, planning, programming

N. Munier, *A Strategy for Using Multicriteria Analysis in Decision-Making: A Guide for Simple and Complex Environmental Projects*, DOI 10.1007/978-94-007-1512-7_8,

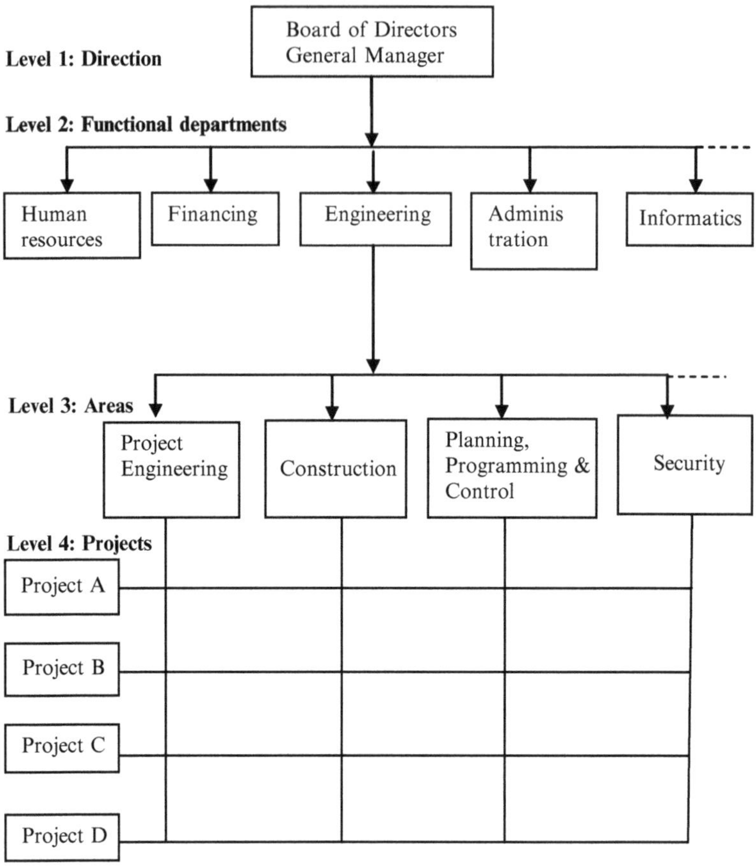

Fig. 8.1 Matrix structure of a consulting firm

and control, etc. Informatics supplies the necessary computer support, as for instance software used for foundation calculations, structural computations, Enterprise Resource Planning, etc. The different projects the firm is undertaking are in Level 3. This is a matrix structure because the functional departments in columns and the diverse projects in rows constitute an interrelations matrix.

Linear Programming works under the same principle (See Fig. 8.2). Here the common functional areas of level 2 are equivalent to selection criteria which are common to all projects. For instance, there are technical, economical, social and environmental issues. However, it is possible that not all projects are associated with criteria. Let us see:

Different projects are in rows at Level 3. Assume that the project consists in building a large industrial plant for car manufacturing, which location could be any of already selected sites in three different countries. The decision-making problem is then to determine the best location for the plant.

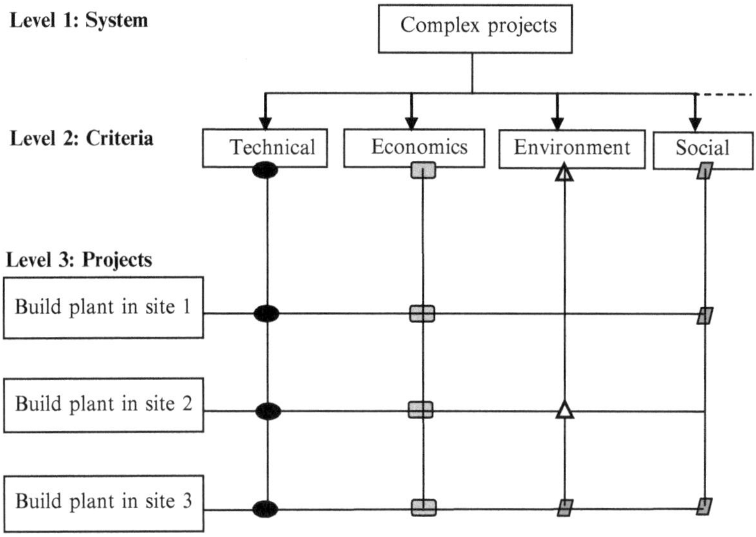

Fig. 8.2 Interrelationship between projects and different evaluation attributes

Technical and economic criteria influence all projects. There are criteria associated with large civil works, or some linked with economic factors, or criteria connected with social issues, or any other circumstance. There are other influences such as observed in site 2, which is affected by an environmental criterion (for instance related with the fact that the site is too close to a natural park and that emissions from the industrial plant could affect the ecology of the area). Most probably, the social criterion should also be present because the necessity of an abundant work force with people from the area, but it must be clear that in most of these cases all criteria participate directly or indirectly in all projects.

These interrelations materialize as has been expressed many times, through numerical values (scores), which show how good a given project fits the demands of a criterion, or how much said project participates in that criterion. For instance, a criterion normally used is availability of capital; thus, at the intersection of this criterion with each one of the alternatives or projects there will be a score showing what the plant would cost at each site. For another criterion such as the minimum generation of energy (from the point of view of economies of scale, because less than a certain output could be not profitable), the score in MW at each intersection will indicate the continuous expected production from each project.

It is clear that if economy were the only criterion to consider it would not be necessary to use multicriteria techniques, for the selection would be for the site that offers the least cost. In reality, this procedure, extensively employed in the past, utilized the cost/benefit analysis where the only important criterion was the economic aspect, normally expressed as Internal Rate of Return, Net Present Value or by the Pay Back period.

Actual cases are much more complex. It could be that regarding cost, site 2 is better than site 1 and 3, but probably this preference will not hold when other criteria are considered. These other criteria could be for instance, availability of workforce, access to harbours, distance to markets, distance to suppliers, the existence or not of suppliers of metal products and mechanical expertise in the area, the existence of schools to provide trained people such as engineers, welders, specialized painters, lathe operators, etc.

There are many other important criteria to take into account. For instance, criteria related with taxes are usually site-dependent, because the local government can offer a reduction in taxes if the plant is located in its jurisdiction, as well as criteria associated with the feasibility of capital repatriation in case of foreign firms investing. Another could be the easiness of air communications by means of an international airport in the area, or labour scenario and labour unions, etc. As appreciated, there are many factors to consider and which materialize in criteria.

SIMUS is applied to solve this type of complex problem, and as any other methodology, the model does not decide, it only steers the attention of the DM in those aspects that are significant and that provide satisfactory solutions. Besides, it supplies elements for the DM to make a judicious judgment, as are marginal values for criteria. The largest advantage of LP is that it permits building of a model, as close as possible to reality, because it is able to work with hundreds of criteria of all type and dimensions. Finally, there are LP applications that can work with uncertain data using fuzzy logic (See Hansen 1996).

8.2 Strategic Planning of Complex Projects Including a Portfolio of Projects – Aspects to Consider

Strategic planning is defined as the preparation and development of plans and programs to reach an objective, sometimes called '*project dimensions*'. It starts with the preparation of the project and to do that, all different projects are analyzed and as much as possible grouped into homogeneous sectors. That is, three projects, the first dealing with a massive flu vaccination program, the second oriented to provide technical training on different trades to people in unemployment, and the third aimed at decreasing transit accidents, could be grouped within the '*Social projects*' area. By the same token, it is possible to group projects for the construction of domestic sewers, storm drains, pavements, opening of new roads, etc., within the '*Infrastructure*' area. Projects such as the construction of an urban incinerator for urban waste, creation of new green spaces for citizens enjoyment, or a new regulation establishing more severe limits to environmental pollution, might be grouped into '*Environment*', and so on.

In strategic planning this grouping occurs, amongst other reasons, in order to have a responsible entity in each area, which is normally the way to execute a work nowadays. That is, a Health Department of a City Hall includes a multitude of projects with a large scope, comprising from hospitals construction to the provision of ambulance services, from analyzing food outlets to the extermination of rodents in sewers.

Normally, each Department has an annual, triennial or a 5-year budget, and with the function to execute all projects, programs and plans on their own and within this budget. This introduces a restriction, since when programming the execution of work in a city its cost must honour the money amount assigned to each department by the City Hall.

Now the process continues with phasing the system. Each project is broken down in parts to facilitate the task of its analysis in detail, but of course, this subdivision does not reach the same level of detail in all projects. For instance, let us analyze a project for technical instruction, where there is the need to define the population target, the nature of practical works, necessities of equipment and tools, etc., and obviously, another project within the system probably will have a different structure. Once this step concludes there will be a very clear picture of what to do, requirements, necessities, etc. for each area and project and besides, it will be possible to identify related projects so they can share resources, or categorize projects, which complement each other.

As an example of ancillary aspects and parts to consider in a complex project, assume the task of determining the best location for hydro projects. It is necessary to identify in detail, beyond the technical factors, other factors such as alteration of native species habitat, fish hatching areas, forests to be flooded, population relocation, etc. For instance, it may be necessary to think about the construction of a *'fish ladder'*[1] to allow trout, to bypass the dam, and swim up-stream to arrive at their hatching areas. This device would have been unnecessary in other locations only a few km up-stream, but that had neither technical nor geological requisites.

The next dimension, the project environment, involves the analysis of each project. For instance, are there sufficient unemployed and interested people to attend the courses above mentioned? Are there suitable places to impart learning? What materials are needed and what will be its cost? Entering in the technical dimension and coming back to our system, which methodology will be used for selection? As explained in Chap. 3 there are several techniques that can be employed, but it should be remembered that in today's standpoint every project must consider simultaneously the economy, the social aspect and the environment, that is the sustainable concept. We can summarize this by saying that a well-designed project lies in the geometric space where these conditions exist.

As for the last dimension, the metaproject, any project requires active participation and communication between people involved in it, i.e. between technical people, the analysts and the DM. The latter establishes the pertinent criteria and delimits them as a function of technical, social, economical and environmental concepts. This includes the methodology to use and the gathering of expert opinions.

[1]Fish ladder: Contrivance that carries water around a dam through a series of stepped baffles or boxes and thus facilitates the migration of fish; also known as fish way. McGraw-Hill Dictionary of Scientific & Technical Terms, 6E, Copyright © 2003 by The McGraw-Hill Companies, Inc.

8.3 Solving Complex Projects with Linear Programming

The objective of this section is to show, obviously very condensed, how complex and very complex projects are solved by Linear Programming. These refer to authentic cases, and, when possible, information is given about the author/s or organism that promoted a project, and technical data.

8.3.1 Case Study: River Basin Planning

Complexity depends on many circumstances; for that reason, probably the best way to proceed is to make a comment on an actual case. The following example examines a hydro project, which, as most river basin projects, is multiobjective by nature, because of the different objectives pursued by the different regions involved.

8.3.1.1 Background Information – General Objective

The objective was to take maximum advantage of the Colorado River (Argentina) water, or in other words, determine the best use of its water. The Department of Civil Engineering at the Massachusetts Institute of Technology (MIT), under contract with the Republic of Argentina, solved the problem by utilizing three models to reach a solution: a screening model, a simulation model, and a sequential model. The first was a linear multiobjective programming model, using some integer variables to determine which projects to build and their appropriate sizes (Cohon 1978).

8.3.1.2 Brief Description

Although it originates high in the Andes range, it is a plain river for nearly 90% of its course. It is located in northern Patagonia, in Argentina, and runs 114 km until draining into the Atlantic Ocean and serving as a natural border for five provinces. The project consisted not only in taking the maximum advantage of a scarce resource (with a modest average flow of 148 m^3/s), but also to satisfy diverse expectations, including decreasing as much as possible its ecological impacts.

8.3.1.3 Conflicting Interests and Restrictions They Generate

Each of the affected provinces had its own plan for an integral hydraulic utilization of the resource, involving consumptive use as in the case of irrigation, with different types of crops, with diverse demands per hectare according to the crop, and for seasonal rotation and diverse water demand. There was a non-consumptive use of water as well, and a dam envisioned to store water and form an artificial lake to be further used to generate electrical energy.

8.3.1.4 Restrictions

Nature's Restrictions

There were restrictions at the mouth of the river especially because of the high salt content, since the water carried salt from upriver soil contamination, increased by residues of pesticides and herbicides. On top of that, there were compensations to pay to the province most affected by salting.

One of the criteria within this scheme required establishing a maximum flow at the mouth of the river to avoid a decrease in the volume of water feeding the aquifers, and then allowing seawater to intrude into them, which of course would diminish the economic benefit that said province would gain from the river.

The analysis established restrictions regarding evaporation when considering alternatives for damming the river, or open aqueducts to transport water to other areas, but mainly devices to act as a defence against flooding in order to protect crops during the de-icing in the Andes. On the other hand, gates in dams could regulate flow in wintertime when the flow is at its lowest. Also considered were serious erosion problems, as well as the construction of defences to protect against erosion.

The river can reach, in a cyclical level every 11 years, 11,000 m^3/s flow, which is more than 70 times its average flow. These figures point to the variability of the flow, which is an aspect to consider in the analysis of the data, and corresponds to the need for meticulous planning.

Naturally, there were more restrictions in this project but those enumerated are enough to show its complexity. This complexity can also be appreciated in the interrelationship between projects or alternatives, as follows:

8.3.1.5 Interrelations Between Alternatives

Excluding, such as in agriculture projects. Since water was scarce there was insufficient quantity to support all the projects, therefore it was necessary to reject some to benefit others, since no simultaneity was possible. Of course, the model did this discrimination.

Complementary, as in the case of the hydroelectric plant and irrigation channels, for if the electric plant is not built there would be no energy for distribution and well pumps. On the other hand, without crops there is no sense in generating electric energy for there would be no use for it.

Dependents, as those related with consumption and aquifer recharging.

Budget restrictions, and spread in several fiscal periods.

Restrictions about different *types of crops* in each location, because crops must rotate in order not to exhaust the soil.

Restrictions emerging from the *physical conditions of the soil in some areas* and that required the construction of certain undertakings.

Political considerations, with five provinces which demanded to have the right to various flows based on the population density in those areas where the river runs

8.3.1.6 Impacts

There were besides some chain impacts. River water is crystalline and very pure at its birthplace high in the Andes. As the river runs through arid areas, it gathers salt and contaminants from crops, carried by scarce rains to the river. Consequently, the river reaches its mouth with high contamination levels, as was already noted above.

Now, less salt contamination at the mouth means diminishing the contamination of up-waters, which is impossible since contamination exists along the river. Since water volume is finite, the only way to provide less contamination at the mouth is to allow salt to be diluted in a greater volume of water which implies restricting the consumptive use of water up-river. The question is where to restrict the use of water in order to proceed with justice and fairness? The answer is far from being an easy one.

If it is true that a simple decision could be to reduce consumption in those undertakings where it is greater, this is not so straightforward, since water consumption also depends on other factors. These factors are the type of crops and the size of hectares to cultivate for the operation to be profitable, which depends on the economy of scale of that crop, which in turn is a function of prices. Besides, it is necessary to take into account that each province has the right to a quota of water and consequently it would not tolerate receiving below the level allowed by legislation and agreement between provinces.

It is necessary to consider that this chain of impacts is an example of only one restriction that is the saline concentration in down-waters. Therefore, it is necessary to take into account that there could be various other restrictions forming a similar chain of effects. Table 8.1a and b[2] is a criteria matrix, which does not faithfully represent the original case, because it has been altered to show more possible difficulties which may come up in a case like this, but this addition does not constitute a problem for its solution.

8.3.1.7 Solution

In reality all the comments made above on this project constitute a modest and general description of this multiobjective problem whose complexity is considerably greater than described here. To appreciate its intricacy, consult Cohon (2003), and Cohon

[2]Tables 8.1a and b are two views of the same table, and it has been partitioned because its length related with the number of projects or alternatives. Thus, Table 8.1a details projects from 1 to 10, while Table 8.1b considers projects from 11 to 20.

et al. (1973), who reported that the project had 629 restrictions or constraints, 665 variables and 8, 0–1 variables. There were also sets of constraints related to Continuity, Reservoirs, Irrigation and Hydroelectric energy. The MIT people solved the problem considering two objectives, which were (a) the minimization of deviation from water previously allocated to regions regarding a regional average user, and (b) the maximization of economic efficiency benefits. The comparison between these two objectives led to a very interesting analysis of the tradeoffs between these objectives.

8.3.1.8 Sensitivity Analysis

> There were 28 final runs of the screening model, which included sensitivity analysis of various economic and physical parameters (Cohon et al. 1973).

8.3.1.9 Comments on This Case

In this example, it is possible to appreciate different characteristics common to complex projects such as:

- Number of intervening elements with 20 different projects or alternatives and pertaining to different areas such as agriculture, farming, mining, hydroelectricity, water transfer and storage and environmental protection, and all of them competing for the same resource: water,
- A series of 32 diverse criteria to be satisfied related with Economy, Financing, Social, Environment, Sustainability, Technical, and Territorial,
- A large series of relationships between projects, such as between crops and farming in different provinces, erosion and flood protection.

The only purpose of this section has been to show the characteristics of a complex project, and the potential of using LP for its solving. Because its importance this project is considered paradigmatic in the technical literature on river basins.

8.3.2 Case Study: Selecting Environmental Indicators

8.3.2.1 Background Information – General Objective

In many activities, there is a need to check periodically the condition or state of something. In environmental matters it is common to use metrics called *'indicators'* which are quantities that measure the state of the environment in its many aspects, and as that, they apply to a lot of areas such as rivers, air, soil, water, forests, ecosystems, wastes, etc. These indicators, when properly selected can give the equivalent of an 'X-ray' picture of the environment, and usually relate to international standards.

Table 8.1 Preparation of a criteria matrix based on effects and impacts from projects

Project I.D.		1	2	3	4	5	6	7	8	9	10
Projects		Crops in Neuquén province	Mining Project in Neuquén province	Crops in Mendoza province	Crops in Río Negro province	Fruit trees in Río Negro province	Hydroelectric plant in Río Negro province	Crops in La Pampa province	Crops in Buenos Aires province	Floods protection	Erosion protection
(a)											
Economic criteria											
IRR	MAX	x		x	x	x	x	x	x	x	
NPV	MAX	x		x	x	x	x	x	x	x	
Electric generation	MAX						x				
Soil valuation	MAX	x		x	x	x		x	x		
Tourism	MAX					x	x				
Financial criteria											
Annual interest for financing	MIN	x		x	x	x	x	x	x		
Money pay back period	MAX	x		x	x	x	x	x	x		
Payment capacity	MAX	x		x	x	x	x	x	x		
Social criteria											
Number of benefited families	MAX				x	x			x		
Increase in per capita income	MAX										
Employment generation	MAX	x		x	x	x		x	x		
Territorial integration	MAX										

Environmental criteria											
Alteration of micro ecosystem	MIN										
Geological faults	MIN						x				
Evapotranspiration level	MIN	x		x	x	x	x	x	x		
Aquifer recharging	MAX				x	x					
Stop desert advancement	MAX			x	x			x			x
Sustainable criteria											
Soil conservation	MAX									x	x
Conservation of wildlife	MAX		x								
Creation of a fish population	MAX										
Promoting biodiversity	MAX	x	x	x	x	x	x	x	x	x	x
Technical criteria											
Accessibility	MAX		x								
Rain Regime	MAX	x	x	x	x	x		x	x	x	x
Permeability	MAX	x	x	x	x	x		x	x	x	
Road slopes	MIN	x	x	x	x	x	x			x	x
Slope for hydroelectric plants	MAX						x				
Transmission lines	MIN										
Territorial criteria											
Neuquén province	MAX	x	x								
Mendoza province	MAX			x							
Río Negro province	MAX				x	x					
La Pampa province	MAX						x				
Buenos Aires province	MAX								x		

(continued)

Table 8.1 (continued)

Project I.D.		11	12	13	14	15	16	17	18	19	20
Projects		Hydroelectric plant in La Pampa province	Hydroelectric plant in La Pampa province plus irrigation	Water reservoir in Río Negro province	Swamps in Buenos Aires province	Potable water distribution in Neuquen province	Potable water distribution in Mendoza province	Rural electrification	Water transfer from Negro River by open channel	Water transfer from Negro River by aqueduct	Necessity to decrease salinity at the mouth
(b)											
Economic criteria											
IRR	MAX	x	x					x		x	
NPV	MAX	x	x					x		x	
Electric generation	MAX	x	x						x	x	
Soil valuation	MAX	x	x						x	x	x
Tourism	MAX	x	x					x			
Financial criteria											
Annual interest for financing	MIN			x	x						
Money pay back period	MAX										
Payment capacity	MAX	x				x	x	x	x	x	
Social Criteria											
Number of benefited families	MAX	x				x	x	x			
Increase in per capita income	MAX					x	x	x			
Employment generation	MAX					x	x	x			x
Territorial integration	MAX							x			

Environmental criteria											
Alteration of micro ecosystem	MIN			x	x				x	x	
Geological faults	MIN					x					
Evapotranspiration level	MIN			x						x	
Aquifer recharging	MAX		x	x							
Stop desert advancement	MAX	x	x						x		
Sustainable criteria											
Soil conservation	MAX	x									
Conservation of wildlife	MAX										
Creation of a fish population	MAX										x
Promoting biodiversity	MAX				x						
Technical criteria											
Accessibility	MAX			x	x						x
Rain Regime	MAX										
Permeability	MAX	x		x	x			x			
Road slopes	MIN										
Slope for hydroelectric plants	MAX			x	x						
Transmission lines	MIN	x				x	x		x		
Territorial criteria											
Neuquén province	MAX							x			
Mendoza province	MAX										
Río Negro province	MAX										
La Pampa province	MAX										
Buenos Aires province	MAX			x							

As an example, actual measures of an indicator such as air quality in a city gives an idea of how good or bad the atmospheric air is. This of course have many implications for human and animal health, vegetation, and even for monuments, when attacked by noxious gases laced with SOx.

In practice, there are so many aspects to control that the number of indicators can amount to hundreds, each one treating a different aspect, and all relevant, although some more significant than others. Naturally, indicators must comply with also a large list of requirements, regarding a true or even an approximate representation of the fact investigated, as well as permanence in time, being understandable for most people, a reasonable collection cost, etc. However, it is materially impossible for anybody to examine and extract conclusions from hundreds of indicators with different units of measure and pertaining to many different aspects, and this is the reason of this section, that is, it aims at determining a short set of core indicators representing as many aspects as possible, and supplying the maximum amount of information.

It is perhaps a different kind of decision-making problem, because it requires identifying, amongst hundreds, which are the most representative indicators to monitor the health of the environment. However, we are not looking here for a ranking but for a manageable list of may be 20 or 25 indicators, to allow us to check the state of the environment and also to monitor the extent of responses to certain correction actions, such as those related to a stricter control of quality of industrial effluents into a river. Indicators are also useful to determine a trend, that is performance of what is measured as a function of time, such as wild life population in an area, or number of day of clean air, etc.

This is the objective of the method proposed in this case using Linear Programming, which by the way, this writer believes is the only tool able to do this, not only because of the sheer size of the problem but perhaps more important, their complexity. Why are these problems complex?

For several reasons, such as:

- Scope, because society uses many ways to contaminate water, and each one produces a different impact and with different intensity and damage For instance if we are concerned with water quality there are several factors related with this issue, such as:
  - Organic wastes mainly from food industry discharges, producing fish migration and death of aquatic species. As an example, quality of discharged effluent can be measured by an indicator called (BOD_5), Biological Oxygen Demand,
  - Nitrogen, Phosphorous and Potassium content, from water treatment plant discharges, favouring algae growth that consumes oxygen needed for aquatic life,
  - Salinity, as seawater intrudes into fresh water, damaging water quality in aquifers, because this impact is sometimes linked to excessive water extraction from the aquifer,
  - Micro-organisms, from fecal coliform bacteria such as those encountered in untreated residential wastewater discharged raw into rivers and to the sea in many places around the world, with extremely dangerous consequences to human life,

- Metals, derived from industrial plants such as plating shops, which are very dangerous and producing different diseases in human beings,
- The presence of herbicides, pesticides and fertilizers, which come from agriculture runoff, contaminating water and favouring algae growth,
- pH level, in bodies of water affected by acid rain, affecting aquatic life,
- Temperature increase, by hot discharges of water used for steam condensing in energy power plants, affecting aquatic life.
- Turbidity possibly linked with diseases-causing micro viruses, etc.

- Links. Most indicators are linked with other in a serial process.
- Most government agencies dealing with this subject need different indicators for certain areas such as Economy, Public Health, Environment, Social, Safety, etc. Complication comes from the fact that every one of these areas requires a certain number of indicators from the final selected list.
- The need to adhere to some environmental framework,[3] such as that developed by the Organization for Economic Co-Operation and Development (OECD).

This example does not reveal for obvious reasons the total complexity of the case, but corresponds to a real scenario developed by this author for the government of a Canadian Province, and consequently, it is only a reduced version of the actual case, but the applied principles are the same. The objective is to select, out of an initial set, a set of indicators, which must provide the maximum information, and at the same time complying with all the requirements expressed by criteria.

8.3.2.2 The Problem

A provincial environmental government agency has 19 indicators (in the actual case there were 54), and wishes to use no more than eight indicators out of the initial 19. These original indicators are subject to six technical criteria plus three criteria linked with operative requirements. These three last criteria have a particular feature: Each one of them demands a minimum percentage of indicators, regarding the total number of final indicators required. This is very important because the DM can fix these percentages, and get exactly the number of indicators required, as will seen in this example. If this circumstance is not considered, it could be that the model assigns most indicators to one or two areas leaving no indicators for others. All data is in Table 8.2.

---

[3]What is a framework for indicators? It is a theoretical structure, used for:

- Systematising data and information,
- Selecting indicators,
- Making explicit the existing interactions between different issues. Several different frameworks have been developed, each one addressing a specific purpose, but probably the most used framework is the OECD approach known as P-S-R, or Pressure, Stress, Response. It was developed to link **P**ressures on the environment by pointing out the human actions or factors (*'stressors'*) that exert pressure on the environment. The **S**tress refers to the effects produced by these stressors, and the **R**esponse is what is being done to correct these effects.

Table 8.2 Areas and indicators

Areas	Aspects and indicators
Human health	Acid rain (SO_2 emissions)
	Climate change (Greenhouse gas emissions)
	Stratospheric ozone (Average annual ozone levels)
	Stratospheric ozone (Concentration)
	Urban air quality (Levels of ground-level ozone)
	Urban air quality (Other air pollutants)
Natural resources	Forestry (Protected forest areas)
	Forestry (Population of forest birds species)
	Forestry (Defoliation)
	Agricultural soils (Residual nitrogen)
	Municipal water use (Consumption per inhabitant)
	Municipal water use (Water used for watering parks)
Transportation	Energy consumption (Fossil fuel consumption)
	Passenger transportation (Passenger travel by mode)
	Passenger transportation (Passenger cars occupancy)
	Passenger transportation (Bicycle paths)
	Municipal solid waste (Per capita disposal)
	Municipal solid waste (Total disposal)
	Annual average of residential waste (Municipal tonnage used for electrical generation)

8.3.2.3 Indicators and Criteria

Original indicators in this example pertain to three main areas (see Table 8.2) (but there could be many more), while Table 8.3 describes indicators. See Table 8.4 with data including weights for indicators, result of a consensus between the DM and experts, and based in a 1–10 scale, the higher the better. Notice that there is not too much discrepancy between indicator weights, which will make the decision even more difficult.

The opinions of three experts allowed establishing scores for an indicator with respect to each of the six technical criteria. These experts independently evaluated each indicator using a 1–10 scale, the higher the better. As it happened, there were indicators for which an expert was unable to make an estimate; in that case no values were reported, and cells identified with (*).

The last column shows the requirements for each criterion or threshold for technical criteria. They come by selecting the smallest score from each row, because each one of them calls for *minimization* action, that is, the result must be above this minimum value. If one criterion would call for a *maximization* action, then the threshold would be the largest score, signalling that the result must be below the largest value. Both actions may blend in the decision matrix, i.e. some criteria can call for maximization while others for minimization.

Criteria 7, 8 and 9 *link with areas* and are elected by the DM. The corresponding *'requirement'* column specifies the number of indicators demanded for each area

Table 8.3 Indicator's description and scope

Aspects	Aspects I.D.	Indicators I.D.	Indicator descriptions
Acid rain	A	1	Sulphur dioxide emissions as a daily average in different parts of the province
Climate change	B	2	Global temperature variations measured as an average around the world
Stratospheric ozone	C	3	Average annual ozone levels as an average of measures in several cities
Stratospheric ozone	D	4	Atmospheric concentrations of ozone-depleting substances at global scale
Urban air quality	E	5	Annual average levels of ground-level ozone in the largest cities of the province
Urban air quality	F	6	Ambient levels of other air pollutants as an average in industrial zones and related with solar activity
Forestry	G	7	Protected forest area in selected forest ecozones in the Eastern of the province
Forestry	H	8	Population status of forest bird species in selected forested ecozones in the Eastern of the province
Forestry	I	9	Consecutive years of spruce budworm defoliation in the Eastern and Northwest of the province
Agricultural soils	J	10	Residual nitrogen in rivers and large fresh water bodies in the province
Municipal water use	K	11	Daily municipal water use in watering parks and urban forests as well as street cleaning
Municipal water use	L	12	Total daily municipal water use for watering parks and urban forests as well as street cleaning
Energy consumption	M	13	Year total fossil fuel consumption in industries, truck, rail and automobile transportation
Passenger transportation	N	14	Annual average passenger travel by mode in medium size and large cities of the province
Passenger transportation	O	15	Annual average of passenger car and vans occupancy commuting daily into cities
Passenger transportation	P	16	Average of kilometres of bicycles paths and number of people riding daily to work
Municipal solid waste	Q	17	Per capita non-hazardous solid waste disposal and recycling/reuse (whole province)
Municipal solid waste	R	18	Annual average of passenger car and vans occupancy commuting daily into cities
Annual average of residential waste	S	19	Municipal tonnage of residential waste being burnt with energy production vs. land filled

Table 8.4 Data from experts – decision table and indicators selected

Selecting environmental indicators

Indicators proposed

AREA	Human health						Natural resources						Transportation							
ASPECTS I.D.	A	B	C	D	E	F	G	H	I	J	K	L	M	N	O	P	Q	R	S	Sum of weights
WEIGHT	8	6	4	9	3	4	7	7	6	2	8	9	9	5	8	5	5	7	9	121
Indicator I.D.	1	2	3	4	5	6	7	8	9	10	11	12	13	14	15	16	17	18	19	
Normalized weight	0.066	0.050	0.033	0.074	0.025	0.033	0.058	0.058	0.050	0.017	0.066	0.074	0.074	0.041	0.066	0.041	0.041	0.058	0.074	

Experts' estimates of each indicator relevance (The larger the better)

CRITERIA LINKED W/ TECH. SELECTION	A	B	C	D	E	F	G	H	I	J	K	L	M	N	O	P	Q	R	S
Alice																			
1) Relevant to environmental sustainability	10	7	10	10	8	6	7	10	10	10	*	10	10	10	*	7	7	7	7
2) Scientifically credible and tested	9	7	8	9	4	4	4	9	9	9	*	9	9	9	*	7	7	7	7
3) Indicator is agreed by researchers	9	8	7	9	7	8	2	7	9	9	*	8	9	9	*	8	8	8	7
4) Relevant at different scales	7	6	5	7	3	2	0	5	7	7	*	7	7	7	*	4	4	6	5
5) Links to policy priorities	9	5	8	9	7	6	7	7	9	9	*	9	8	10	*	7	7	5	7
6) Relationship to human activity	9	8	8	6	6	5	0	7	9	9	*	8	9	9	*	7	7	8	7
Daniel																			
1) Relevant to environmental sustainability	9	*	9	9	9	8	9	7	7	7	9	*	8	10	9	*	7	8	7
2) Scientifically credible and tested	8	*	8	7	7	8	7	7	7	6	8	*	8	9	8	*	8	8	7
3) Indicator is agreed by researchers	8	*	7	8	8	4	8	7	7	8	7	*	7	9	7	*	8	7	7
4) Relevant at different scales	7	*	6	6	6	6	8	5	5	5	7	*	6	7	6	*	6	5	5
5) Links to policy priorities	8	*	8	8	6	9	7	7	7	7	9	*	6	10	6	*	5	5	7
6) Relationship to human activity	8	*	9	8	8	7	8	7	7	6	8	*	9	9	8	*	8	8	7

	Michelle																		
1) Relevant to environmental sustainability	7	7	7	9	9	9	9	9	8	8	7	7	*	8	9	8	8	8	*
2) Scientifically credible and tested	7	8	7	8	8	6	8	7	7	8	7	7	*	8	8	8	6	7	*
3) Indicator is agreed by researchers	7	7	7	8	8	7	8	8	8	8	8	8	*	8	8	8	7	8	*
4) Relevant at different scales	5	3	5	3	7	6	6	6	6	6	6	6	*	5	5	5	6	6	*
5) Links to policy priorities	7	5	7	7	8	6	8	8	6	7	5	5	*	6	6	6	5	5	*
6) Relationship to human activity	7	6	7	7	8	8	7	7	7	7	8	8	*	8	8	8	8	8	*

(*) No estimate available for this indicator

	Average values																			Results from computation	Action	Requirements
1) Relevant to environmental sustainability	8.67	7.00	8.67	9.33	8.67	7.67	8.33	8.67	8.33	8.33	8.00	5.67	9.00	9.33	9.00	7.50	7.33	7.67	7.00	66.67	≥	5.67
2) Scientifically credible and tested	8.33	7.50	7.67	8.00	6.33	6.00	6.33	7.67	7.67	7.67	7.50	5.33	8.50	8.67	8.00	7.50	7.00	7.33	7.00	59.67	≥	5.33
3) Indicator is agreed by researchers	8.00	7.50	7.00	8.33	7.67	6.33	6.00	7.33	8.00	8.33	7.50	5.33	8.00	8.67	7.50	8.00	7.67	7.67	7.00	58.00	≥	5.33
4) Relevant at different scales	6.67	4.50	5.33	5.33	5.33	4.67	4.67	5.33	6.00	6.00	6.50	4.33	6.50	6.33	5.50	4.50	5.33	5.67	5.00	44.83	≥	4.33
5) Links to policy priorities	7.67	5.00	7.67	8.00	7.00	7.00	7.33	7.33	7.33	7.67	7.00	4.67	7.00	8.67	6.00	6.50	5.67	5.00	7.00	55.00	≥	4.67
6) Relationship to human activity	8.00	7.00	8.00	7.00	7.33	6.67	5.00	7.00	7.67	7.33	8.00	5.33	9.00	8.67	8.00	7.50	7.67	8.00	7.00	57.33	≥	5.00
Forcing all indicators to be considered	1	1	1	1	1	1	1	1	1	1	1	1	1	1	1	1	1	1	1	8	=	8
CRITERIA LINKED WITH AREAS																						
(Minimum percentage of indicators per area)																						
7) Human health	1	1	1	1	1	1														2.00	≥	2.00
8) Natural resources use							1	1	1	1	1	1								4.00	≥	4.00
9) Transportation													1	1	1	1	1	1	1	2.00	≥	2.00

Indicate here number of indicators required (→ 8)

Indicator I.D.	1	2	3	4	5	6	7	8	9	10	11	12	13	14	15	16	17	18	19
Indicators selected	1	0	0	1	0	0	1	1	0	0	1	1	1	0	1	0	0	0	0

Objective function: 0.537

See here number of indicators produced: 8

Percentage obtained of indicators for HUMAN HEALTH area	0.25
Percentage obtained of indicators for NATURAL RESOURCES area	0.50
Percentage obtained of indicators for TRANSPORTATION area	0.25

(they can also be expressed as percentages of the total number of indicators required, in this case, eight, which is boxed). The DM fixes these requirements or thresholds, and in this case, he/she has deemed that the most important is '*Natural resources use*' and has assigned to it 50% of the eight final indicators and 25% to each one of the other two (that is, 4, 2 and 2). Naturally, the DM can choose whatever mix of percentages he/she considers most appropriate. At the right end of row '*Number of indicators required*' is (boxed) the location where the DM places the number of final indicators he/she requires.

The objective function (Z=0.537) calls for maximization; it is in a boxed cell at the bottom left and expresses,(but not seen) the sum of products between indicators' weights and results displayed in the '*Indicators selected*' row. As one can see, in this row there is a series of '1 s' indicating the eight selected indicators, which are:

1. Sulphur dioxide emissions as a daily average in different parts of the province,
4. Atmospheric concentrations of ozone-depleting substances at global scale,
7. Protected forest area in selected forest ecozones in the Eastern of the province,
8. Population status of forest bird species in selected forested ecozones in the Eastern of the province,
11. Daily municipal water use in watering parks and urban forests as well as street cleaning,
12. Total daily municipal water use for watering parks and urban forests as well as street cleaning,
13. Year total fossil fuel consumption in industries, truck, rail and automobile transportation,
15. Annual average of passenger car and vans occupancy commuting daily into cities.

8.3.2.4 Other More Down-to-Earth Scenario

This already analyzed example was worked under the assumption that scores for all indicators and for criteria correspond to evaluations from the three experts, but this is rarely the case, because often the experts do not have enough information to make an appraisal on certain indicators, as was assumed in this example. This new case belongs to the same example but considers that many more scores are unknown because of lack of information, corrupt or incomplete data, uncertainties, or other reasons.

This more realistic picture is displayed in Table 8.5. Observe that many indicators do not have scores in some criteria from the three experts; for this reason, and also because the average values do not reflect the experts' opinions in full, another procedure applies.

It consists in replacing score values by '1 s' when there is an average value, whatever it might be. It practice, it could also be that experts do not mutually agree that there is a relationship between an indicator and a certain criterion in particular. This fact is reflected for instance in indicator number 4 where Alice and Michelle

Table 8.5 Situation with incomplete data from experts – decision table and indicators selected

Selecting environmental indicators

Indicators proposed

AREA	Human Health						Natural Resources						Transportation							
ASPECTS I.D.	A	B	C	D	E	F	G	H	I	J	K	L	M	N	O	P	Q	R	S	Sum of weights
WEIGHT	8	6	4	9	3	4	7	7	6	2	8	9	9	5	8	5	5	7	9	121
Indicator I.D.	1	2	3	4	5	6	7	8	9	10	11	12	13	14	15	16	17	18	19	
Normalized weight	0.066	0.050	0.033	0.074	0.025	0.033	0.058	0.058	0.050	0.017	0.066	0.074	0.074	0.041	0.066	0.041	0.041	0.058	0.074	

Experts' estimates of each indicator relevance (The larger the better)

CRITERIA LINKED W/ TECH. SELECTION	A	B	C	D	E	F	G	H	I	J	K	L	M	N	O	P	Q	R	S
Alice																			
1) Relevant to environmental sustainability				9					9									7	
2) Scientifically credible and tested			7				2						9						
3) Indicator is agreed by researchers					3					7									
4) Relevant at different scales		5					7					9		10		7			7
5) Links to policy priorities						5		7					9	9				8	
6) Relationship to human activity				8															
Daniel																			
1) Relevant to environmental sustainability				7					7									8	
2) Scientifically credible and tested			7				8				7		7		7				
3) Indicator is agreed by researchers					6					5									
4) Relevant at different scales							7					8		10					7
5) Links to policy priorities		7				7		7					9	9				8	
6) Relationship to human activity																			
Michelle																			
1) Relevant to environmental sustainability				8					7									7	
2) Scientifically credible and tested			7				8				8				8				
3) Indicator is agreed by researchers					7					6									
4) Relevant at different scales		5										5		6		6			
5) Links to policy priorities						8		7						8				8	
6) Relationship to human activity				7					7	7				8				6	

(continued)

Table 8.5 (continued)

Average values

ASPECTS I.D.	A	B	C	D	E	F	G	H	I	J	K	L	M	N	O	P	Q	R	S
Indicator I.D.	1	2	3	4	5	6	7	8	9	10	11	12	13	14	15	16	17	18	19
1) Relevant to environmental sustainability				8.00					7.67									4.67	
2) Scientifically credible and tested			7.00				6.00				7.50		8.00		7.50				
3) Indicator is agreed by researchers					5.33					6.00									
4) Relevant at different scales		3.33					7.00					7.33		8.67		6.50			7.00
5) Links to policy priorities		7				6.67		7.00					9.00	8.67				8.00	
6) Relationship to human activity				7.50					7.00	7.00				8					

Identification of indicators with values and those the DM wants to be considered

	A	B	C	D	E	F	G	H	I	J	K	L	M	N	O	P	Q	R	S	Results from computation	Action	Requirements
1) Relevant to environmental sustainability	1			1					1									1		3	≥	3
2) Scientifically credible and tested			1				1				1		1		1					2	≥	2
3) Indicator is agreed by researchers					1					1										2	≥	2
4) Relevant at different scales		1					1					1		1		1			1	1	≥	1
5) Links to policy priorities		1				1		1					1	1				1		2	≥	2
6) Relationship to human activity				1					1	1				1			1			2	≥	2
Forcing all indicators to be considered	1	1	1	1	1	1	1	1	1	1	1	1	1	1	1	1	1	1	1	8	≥	8
CRITERIA LINKED WITH AREAS (Minimum percentage of indicators per area)																						
7) Human health	1	1	1	1	1	1														6	≥	2
8) Natural resources use							1	1	1	1	1	1								6	≥	4
9) Transportation													1	1	1	1	1	1	1	6	≥	2

Indicate here number of indicators required → 8

Indicator I.D.	1	2	3	4	5	6	7	8	9	10	11	12	13	14	15	16	17	18	19
Indicators selected	1			1	1					1	1		1					1	1

Objective function: 0.455

See here number of indicators produced → 8

Table 8.5 *Situation with incomplete data from experts - Decision table and indicators selected*

believe that it is related to criterion *"Relationship with human activity"* (shaded cell), while Daniel does not share this opinion (broken line cell).

Observe that there are two indicators without scores (1 and 17) because again, the experts were probably not confident in estimating them, or for whatever reasons. Nevertheless, the DM wants them to be taken into account and linked with some specific criteria. Thus, *'Acid rain'* indicator (# 1) is associated to criterion *'Relevant to environmental sustainability'* (shaded and broken line cell), and indicator *'Municipal solid waste'* (# 17) (broken line cell), will relate to criterion *'Relationship with human activity' both* using '1 s'.

Thresholds in column *'Requirements'* are now a DM decision, since they are subject to what the DM demands regarding quantities of indicators per criterion. For instance, he/she can think this way *'Because environmental sustainability* (criterion number 1), *is so important I want at least three indicators related with this criterion'*. This is specified in the same row and in column *'Requirements*', and similarly for other criteria. Running the model again gets a new result, and can watch that all requirements are honoured. Notice that the model in some cases, as in the Area criteria, produces more indicators than required (remember that the requirement is *'at least'*), and there are more criteria (6, 6, 6) (from column *'Results from computation')* than those originally established (2, 4, 2).

The objective function is (Z = 0.455) and calls for maximization, and the selected indicators, shown in the *'Indicators selected*' are:

1. Sulphur dioxide emissions as a daily average in different parts of the province,
4. Atmospheric concentrations of ozone-depleting substances at global scale,
5. Annual average levels of ground-level ozone in largest cities of the province,
10. Residual nitrogen in rivers and large fresh water bodies in the province,
11. Daily municipal water use in watering parks and urban forests as well as street cleaning,
13. Year total fossil fuel consumption in industries, truck, rail and automobile transportation,
18. Annual average of passenger car and vans occupancy commuting daily into cities,
19. Total non-hazardous solid waste disposal and recycling/reuse (whole province).

As expected, the series of indicators selected is different from the first scenario. As a sort of checking, the reader can verify that, considering for instance criterion number 1 which calls for three indicators (look at the *'Requirements'* column), there are certainly three indicators related with it (look at *'Results from computation'* column.) These three indicators are identified with shaded cells.

8.3.2.5 Changing the Objective Function

Until now, the objective function has consisted in maximizing indicators' weights. However, there is another scenario, which can be more interesting and useful; for instance, what if our objective is to get the maximum quantity of information from the original set of indicators? This can be accomplished using entropy, as mentioned in Sect. 2.5.4 and applied here for this purpose. This is the procedure (see Table 8.6).

1. Refer to *'Original weights'* row (at the bottom of table). These are the same weights used in the two precedent cases.
2. Normalize weights and place the new values in a new row called *'Normalized weights.*
3. Find for each cell of this last row, the product of its value times its log. For instance for indicators number 1 it will be $0.066 \times \log(0.066) = -0.078$; place these new values in a new row called *'(Normalized weight)* × *(log of normalized weight).*
4. Get the ***'Entropy'*** of the system, using this formula: $S = -\sum Normalized\,weight \times \log\,(normalized\ weight)$, which in this case is 1.253.
5. Multiply each normalized value by this entropy value and place them in the last row called *('Total entropy')* × *(Normalized weight').*
6. Use these values for the objective function in cell *'Objective function'*
7. Run Solver and the new results will be displayed in 'Indicators selected' row.

Probably this procedure yields better results, since the model will be utilizing for the objective function new scores found under the proviso of producing the *largest amount of information,* instead of using weight importance as scores, which are subjective. Check that, even if the model selected the same indicators, the objective function changed.

Without entropy the objective function depicted in Table 8.5 is $Z = 0.455$.

Using entropy the value of the objective function depicted in Table 8.6 is $Z = 0.569$.

Consequently, working with this last series of final indicators gives more information on the environment state.

8.3.2.6 Sensitivity Analysis

The model is adequate to perform sensitivity analysis. For instance:

Assume now that the DM wants to choose more than eight final indicators, or a lesser number. This is easy to do, just by placing the new value in the cell indicated by arrow *('Indicate here number of indicators required')* in any of the three tables mentioned. However, a word of caution: It could very well be that the Solver informs us that there is no solution; this is possible if for instance six final indicators are required. The reason is that six final indicators are probably not enough to comply with all restrictions, or inversely, if the DM is asking for too many final indicators in relationship with total number in the original list, the Solver will so inform us.

Table 8.6 Same data from experts but using entropy – decision table and indicators selected

Selecting environmental indicators

Indicators proposed

AREA	Human Health						Natural Resources						Transportation							
ASPECTS I.D.	A	B	C	D	E	F	G	H	I	J	K	L	M	N	O	P	Q	R	S	Sum of weights
WEIGHT	8	6	4	9	3	4	7	7	6	2	8	9	9	5	8	5	5	7	9	121
Indicator I.D.	1	2	3	4	5	6	7	8	9	10	11	12	13	14	15	16	17	18	19	
Normalized weight	0.066	0.050	0.033	0.074	0.025	0.033	0.058	0.058	0.050	0.017	0.066	0.074	0.074	0.041	0.066	0.041	0.041	0.058	0.074	

Experts' estimates of each indicator relevance (The larger the better)

CRITERIA LINKED W/ TECH. SELECTION	A	B	C	D	E	F	G	H	I	J	K	L	M	N	O	P	Q	R	S
Alice																			
1) Relevant to environmental sustainability				9					9									7	
2) Scientifically credible and tested			7				2						9						
3) Indicator is agreed by researchers					3					7									
4) Relevant at different scales		5					7					9		10		7			7
5) Links to policy priorities						5		7					9	9				8	
6) Relationship to human activity				8															
Daniel																			
1) Relevant to environmental sustainability				7					7									8	
2) Scientifically credible and tested			7				8				7		7		7				
3) Indicator is agreed by researchers					6					5									
4) Relevant at different scales							7					8		10					7
5) Links to policy priorities		7				7		7					9	9				8	
6) Relationship to human activity																			
Michelle																			
1) Relevant to environmental sustainability				8					7									7	
2) Scientifically credible and tested			7				8				8				8				
3) Indicator is agreed by researchers					7					6									
4) Relevant at different scales		5										5		6		6			
5) Links to policy priorities						8		7						8				8	
6) Relationship to human activity				7					7	7				8			6		

(continued)

Table 8.6 (continued)

	Average values																					
ASPECTS I.D.	A	B	C	D	E	F	G	H	I	J	K	L	M	N	O	P	Q	R	S			
Indicator I.D.	1	2	3	4	5	6	7	8	9	10	11	12	13	14	15	16	17	18	19			
1) Relevant to environmental sustainability				8.00					7.67									4.67				
2) Scientifically credible and tested			7.00				6.00				7.50		8.00		7.50							
3) Indicator is agreed by researchers					5.33					6.00												
4) Relevant at different scales		3.33					7.00					7.33		8.67		6.50			7.00			
5) Links to policy priorities		7				6.67		7.00					9.00	8.67				8.00				
6) Relationship to human activity				7.50					7.00	7.00				8								
	Identification of indicators with values and those the DM wants to be considered																			Results from computation		Action Requirement
1) Relevant to environmental sustainability	1			1					1									1		3	≥	3
2) Scientifically credible and tested			1				1				1		1		1					2	≥	2
3) Indicator is agreed by researchers					1					1										2	≥	2
4) Relevant at different scales		1					1					1		1		1			1	1	≥	1
5) Links to policy priorities		1				1		1					1	1				1		2	≥	2
6) Relationship to human activity				1					1	1				1			1			2	≥	2
Forcing all indicators to be considered	1	1	1	1	1	1	1	1	1	1	1	1	1	1	1	1	1	1	1	8	=	8
CRITERIA LINKED WITH AREAS (Min.																						
(Minimum percentage of indicators per area)														Indicate here number of indicators <u>required</u>								
7) Human health	1	1	1	1	1	1														6	≥	2
8) Natural resources use							1	1	1	1	1	1								6	≥	4
9) Transportation													1	1	1	1	1	1	1	7	≥	2
Indicator I.D.	1	2	3	4	5	6	7	8	9	10	11	12	13	14	15	16	17	18	19			
Indicators selected	1			1	1					1		1	1		1			1				8
Objective function 0.569													See here number of indicators <u>produced</u>									
Original weights	8	6	4	9	3	4	7	7	6	2	8	9	9	5	8	5	5	7	9	121		
Normalized weights	0.066	0.050	0.033	0.074	0.025	0.033	0.058	0.058	0.050	0.017	0.066	0.074	0.074	0.041	0.066	0.041	0.041	0.058	0.074			
(Norm.weight) x (log of norm. weight)	-0.078	-0.065	-0.049	-0.084	-0.040	-0.049	-0.072	-0.072	-0.065	-0.029	-0.078	-0.084	-0.084	-0.057	-0.078	-0.057	-0.057	-0.072	-0.084	1.253	Total entropy	
(Total entropy) x (Normalized weight)	0.083	0.062	0.041	0.093	0.031	0.041	0.072	0.072	0.062	0.021	0.083	0.093	0.093	0.052	0.083	0.052	0.052	0.072	0.093			
	Scores used for objective function																					

Therefore, it appears that there is a relationship between number of original indicators and number of final indicators required. The explanation lies in the fact that the larger the original number of indicators the better, since the model will have more data to work with. The no-solution could also obey another reason; it could happen that in reality a feasible solution does not exist, because the set of indicators is not fulfilling all the restrictions. In this case, the PM may exert judgement and choose fewer or else additional final indicators, and can also build a table for a different number of final indicators requested, and select one of the options.

8.3.2.7 Comments on This Case

As can be appreciated, this is a complex case not only because of the number of original indicators, which can be hundreds, but too because of all the strong conditions imposed that specify at least a minimum number of indicators per criterion; and similarly there are relationships between area criteria and number of indicators. The example demonstrates the possibility of using the model in several different scenarios. It has also been introduced here the concept of maximizing the objective function with the goal of getting the largest amount of information from the initial universe of indicators. To this writer's knowledge, the use of entropy in a linear programming model for the maximization of an objective function constitutes an innovative approach, albeit largely based on the works of Shannon (1948) and Zeleny (1973).

8.3.3 Case Study: Urban Development

8.3.3.1 Background Information – General Objective

An urban entrepreneur has purchased a tract of land in the periphery of a city for a housing development. The project calls for construction of three types of dwellings, which differ in plot size, floor space and cost to build. Table 8.7 shows the available data. *'Requirements'* column displays different requirements or conditions, while *'Results from computation'* column shows the values obtained by the model for each criterion, that is, it informs us how well the requirement is matched.

Tract of Land

It has a size of 10,000 m^2. Out of it 2,300 m^2 are reserved to streets, sidewalks and green areas. For this reason a requirement cell indicates the maximum available size of land for construction.

Table 8.7 Decision table and solution

Urban development

	Plan for type 1 house	Plan for type 2 house	Plan for type 3 house	Results from computation	Units	Action	Requirements
Investment per dwelling (Euros):	16,800	22,000	26,500				
Area of each lot (m^2)	69	122	160	7,480	m^2	≤	7,700
Minimum number of units of type 1 houses	1			30	dwellings	≥	30
Maximum number of units of type 1 houses	1			30	dwellings	≤	40
Minimum number of units of type 2 houses		1		35	dwellings	≥	35
Maximum number of units of type 2 houses		1		35	dwellings	≤	48
Minimum number of units of type 3 houses			1	7	dwellings	≥	5
Maximum number of units of type 3 houses			1	7	dwellings	≤	10
Minimum number of dwellings	1	1	1	72	dwellings	≥	50
Maximum number of dwellings	1	1	1	72	dwellings	≤	95
Maximum total investment (Euros/ha)	1	1	1	1,462,813	Euros/ha	≤	1,500,000
Minimum total investment (Euros/ha)	1	1	1	1,462,813	Euros/ha	≥	1,250,000
Maximum floor space area (m^2/house)	55	65	95	4,602	m^2	≤	5,000
Minimum floor space (m^2/house)	47	52	80	3,800	m^2	≥	3,800
Maximum density (persons/house)	4.55	5.52	6.30	375	persons	≤	400
Minimum density (persons/house)	4.00	5.20	6.15	346	persons	≥	340
Water consumption (liters/day-house)	1,297	1,573	1,796	106,757	liters/day	≤	115,000
Sewage production baths & kitchens (m^3/day-house	0.98	1.19	1.36	81	m^3/day	≤	90
Electric energy consumption (kWh/day-house)	27.30	33.12	37.80	2,248	kWh/day	≤	3,000
Total cost (Euros) 1,462,813							
Number of dwellings to be built by hectare	30	35	7	72	Total number of houses		

Physical Characteristics of Houses

- Different investments cost for each house,
- Area or size of each plot,
- Maximum and minimum floor space for each house.

Expected Demand

A survey was carried out to determine the profile of demand together with an aggressive advertisement campaign, which yielded the expected demand for each type of house. There is a minimum and a maximum demand number for each type, as follows:

Potential demand for Type 1 house,
Potential demand for Type 2 house,
Potential demand for Type 3 house.

There are rows indicating the estimated *minimum* number of dwellings to build for each house size, from an educated guess made by the entrepreneur based on tract size, approximated demands as per survey results, and economies of scale. Not all houses have the same services and utilities; for instance most priced houses come with Jacuzzi, a large fridge, air conditioning units, clothes washing machines, etc., as well as a considerable large plot size, however, few of these advantages also come with the other size of houses. Another row shows the estimated *maximum* number of dwellings to build, based on the same principles commented on above.

Consequently, for each type there is a range regarding the number of houses to build, and thus for Type 3 for instance, it varies between five as a minimum and ten as a maximum. Considering this interval the Solver computes the most adequate number of units, taking into account all the other restrictions. As can be seen in column *'Results from computation'*, the model finds that the best solution is to build seven units of Type 3.

Floor Space

It has been computed considering an international standard of m^2 of floor space per resident. Consequently, other rows depict the maximum and minimum density, which was compared and adjusted with survey findings.

Basic Services

Estimated water consumption per type of house is shown in liters, calculated according to local standards per person/day, and a total estimate computed, including garden watering. The entrepreneur has consulted City Hall – responsible for water distribution – and he has received the green light for this consumption calculation, otherwise his plans would have not been accepted. The same for total of maximum volume in m^3 of waste water production, and the estimate of electric energy, also agreed with the electric company, which will install a transformer for the development.

Construction Budget

Because of sales uncertainty, the entrepreneur has established limits for his investment, with a minimum of 1,250,000 Euros (as per economies of scale and expected revenue), and a maximum of 1,500,000 Euros (as per maximum banking loans).

The objective, which is to *minimize investment costs,* (in Euros), is boxed in *'Total cost'* row. As mentioned, the Solver takes all data from Excel, processes them, and delivers results back to the spreadsheet, in a row indicated by user and labeled *'Number of dwellings to be built by hectare'*. In this example the result is the following construction program:

Type 1 dwellings: 30,
Type 2 dwellings: 35,
Type 3 dwellings: 7,
Total investment: 1,462,813 Euros.

8.3.3.2 Sensitivity Analysis

In order to find out more about this venture the entrepreneur wishes to learn a couple of things about his project, and to analyze different scenarios, as follows:

1. He demands to know which are the aspects or criteria that most influence his costs.
   This is an easy question for the Linear Programming model as seen in Sect. 6.5, since each time a Linear Programming model is solved, a second 'mirrored' solution of the original problem which is called the *'dual'* is automatically generated. In this example, the dual values obtained from Solver's *sensitivity report* (albeit not shown here), are:

   Minimum number of units of type 1 houses: 1,231,
   Minimum number of units of type 2 houses: 4,775,

   Minimum floor space: 331.

   What do these numbers mean?

   They are the marginal costs of respective criteria. That is, because the problem calls for minimization of costs, each number reflects the increment of costs (objective function) for each unit increase in the criterion. Thus, an increase from 35 to 36 units of type 2 houses, will increment total cost of 4,775

Euros, and the objective function will go from the original 1,462,813 Euros to 1,467,588 Euros (increase of 0.33%). The same analysis is also valid for the other criteria.

Do these numbers tell the entrepreneur that *'Minimum number of units of type 2 houses'* is the most important criterion to examine in order to reduce costs, followed by *'Minimum number of units of type 1 house'* and by *'Minimum floor space'?*

Not really; the statement is true for the two criteria regarding **number** of units of houses, but it does not apply to the third criterion, since this one is not comparable with the others because its unit of measure is different **(m^2)**. In fact, if we increase by only 1 m^2 the minimum limit for *'Minimum floor space'* to 3,801 m^2, in the corresponding requirements cell (boxed), (Table 8.8), the objective function increases to 1,463,144 Euros (increase of 0.023%), albeit the solution does not change.

2. The DM wishes now to analyze what happens if the floor space for Type 2 is reduced for instance in 3 m^2 (from 52 to 49, boxed cell), and in 2 m^2 for Type 1 (from 47 to 45, boxed cell) (Table 8.9). Naturally, in this case there will be a decrease in construction costs, for both types, which the DM estimates at 3,000 Euros for Type 2, decreasing the total cost to 19,000 Euros (boxed cell), and a decrease of 1,300 Euros for Type 1, with a final cost of 15,500 Euros (boxed cell). The model is executed again, which result is 1,379,987 Euros, which is less than the original 1.462,813, and building 77 houses instead of 72.
3. Assume now, that the DM decides to lower the minimum number of units of type 2–30 (instead of 35), (boxed cell in Table 8.10).
4. The model is run again and the result is this:

Now	Before
Type 1 dwellings: 34,	Type 1 dwellings: 30,
Type 2 dwellings: 30,	Type 2 dwellings: 35,
Type 3 dwellings: 10.	Type 3 dwellings: 7.
Total: 74.	Total: 72.

The total cost is now 1,362,000 Euros which is inferior to 100,813 Euros compared with the original of 1,462,813 Euros. Thus, the fact of reducing the lower limit of threshold produces an increase of 2 houses to be built, and considerable savings. One wonders why?

The entrepreneur prepared Table 8.11 and finds that Type 2 houses have the largest cost per square meter of floor space, when compared with the other two options; therefore, using the lower limit the model assigns the minimum quantity to Type 2 houses and increments Type 1 and 3.

5. Suppose now that City Hall has informed the entrepreneur that there was a mistake regarding water supply, and that they will be in no condition to guarantee water service in the specified quantity especially in September, October and November because those are dry months and their reservoir will be at its minimum capacity or nearby. They say that the maximum amount they can guarantee is

Table 8.8 Data with increment of 1 m² to the minimum floor space per house to show sensitivity of model

Urban development

	Plan for type 1 house	Plan for type 2 house	Plan for type 3 house	Results from computation	Units	Action	Requirements
Investment per dwelling (Euros):	16,800	22,000	26,500				
Area of each lot (m^2)	69	122	160	7,482	m^2	≤	7,700
Minimum number of units of type 1 houses	1			30	dwellings	≥	30
Maximum number of units of type 1 houses	1			30	dwellings	≤	40
Minimum number of units of type 2 houses		1		35	dwellings	≥	35
Maximum number of units of type 2 houses		1		35	dwellings	≤	48
Minimum number of units of type 3 houses			1	7	dwellings	≥	5
Maximum number of units of type 3 houses			1	7	dwellings	≤	10
Minimum number of dwellings	1	1	1	72	dwellings	≥	50
Maximum number of dwellings	1	1	1	72	dwellings	≤	95
Maximum total investment (Euros/ha)	1	1	1	1,463,144	Euros/ha	≤	1,500,000
Minimum total investment (Euros/ha)	1	1	1	1,463,144	Euros/ha	≥	1,250,000
Maximum floor space area (m^2/house)	55	65	95	4,603	m^2	≤	5,000
Minimum floor space (m^2/house)	47	52	80	3,801	m^2	≥	3,801
Maximum density (persons/house)	4.55	5.52	6.30	375	persons	≤	400
Minimum density (persons/house)	4.00	5.20	6.15	346	persons	≥	340
Water consumption (liters/day-house)	1,297	1,573	1,796	106,780	liters/day	≤	115,000
Sewage production baths & kitchens (m^3/day-house)	0.98	1.19	1.36	81	m^3/day	≤	90
Electric energy consumption (kWh/day-house)	27.30	33.12	37.80	2,248	kWh/day	≤	3,000
Total cost (Euros) 1,463,144							
Number of dwellings to be built by hectare	30	35	7	72	Total number of houses		

Table 8.9 Data reducing minimum floor space for type 1 and 2 houses, (from 47 m² to 45 m², and from 52 m² to 49 m²), respectively Also changed investment from 16,800 Euros to 15,500 Euros in type 1, and from 22,000 Euros to 19,000 Euros in type 2

Urban development

	Plan for type 1 house	Plan for type 2 house	Plan for type 3 house	Results from computation	Units	Action	Requirements
Investment per dwelling (Euros):	15,500	19,000	26,500				
Area of each lot (m^2)	69	122	160	7,700	m^2	≤	7,700
Minimum number of units of type 1 houses	1			36	dwellings	≥	30
Maximum number of units of type 1 houses	1			36	dwellings	≤	40
Minimum number of units of type 2 houses		1		36	dwellings	≥	35
Maximum number of units of type 2 houses		1		36	dwellings	≤	48
Minimum number of units of type 3 houses			1	5	dwellings	≥	5
Maximum number of units of type 3 houses			1	5	dwellings	≤	10
Minimum number of dwellings	1	1	1	77	dwellings	≥	50
Maximum number of dwellings	1	1	1	77	dwellings	≤	95
Maximum total investment (Euros/ha)	1	1	1	1,379,987	Euros/ha	≤	1,500,000
Minimum total investment (Euros/ha)	1	1	1	1,379,987	Euros/ha	≥	1,250,000
Maximum floor space area (m^2/house)	55	65	95	4,814	m^2	≤	5,000
Minimum floor space (m^2/house)	45	49	80	3,800	m^2	≥	3,800
Maximum density (persons/house)	4.55	5.52	6.30	396	persons	≤	400
Minimum density (persons/house)	4.00	5.20	6.15	363	persons	≥	340
Water consumption (liters/day-house)	1,297	1,573	1,796	112,755	liters/day	≤	115,000
Sewage production baths & kitchens (m^3/day-house)	0.98	1.19	1.36	85	m^3/day	≤	90
Electric energy consumption (kWh/day-house)	27.30	33.12	37.80	2,374	kWh/day	≤	3,000
Total cost (Euros) 1,379,987							
Number of dwellings to be built by hectare	36	36	5	77	Total number of houses		

Table 8.10 Data reducing number of type 2 houses, from 35 units to 30 units

Urban development	Plan for type 1 house	Plan for type 2 house	Plan for type 3 house	Results from computation	Units	Action	Requirements
Investment per dwelling (Euros):	15,500	19,000	26,500				
Area of each lot (m^2)	69	122	160	7,606	m^2	≤	7,700
Minimum number of units of type 1 houses	1			34	dwellings	≥	30
Maximum number of units of type 1 houses	1			34	dwellings	≤	40
Minimum number of units of type 2 houses		1		30	dwellings	≥	30
Maximum number of units of type 2 houses		1		30	dwellings	≤	48
Minimum number of units of type 3 houses			1	10	dwellings	≥	5
Maximum number of units of type 3 houses			1	10	dwellings	≤	10
Minimum number of dwellings	1	1	1	74	dwellings	≥	50
Maximum number of dwellings	1	1	1	74	dwellings	≤	95
Maximum total investment (Euros/ha)	1	1	1	1,362,000	Euros/ha	≤	1,500,000
Minimum total investment (Euros/ha)	1	1	1	1,362,000	Euros/ha	≥	1,250,000
Maximum floor space area (m^2/house)	55	65	95	4,770	m^2	≤	5,000
Minimum floor space (m^2/house)	45	49	80	3,800	m^2	≥	3,800
Maximum density (persons/house)	4.55	5.52	6.30	383	persons	≤	400
Minimum density (persons/house)	4.00	5.20	6.15	354	persons	≥	340
Water consumption (liters/day-house)	1,297	1,573	1,796	109,241	liters/day	≤	115,000
Sewage production baths & kitchens (m^3/day-house)	0.98	1.19	1.36	83	m^3/day	≤	90
Electric energy consumption (kWh/day-house)	27.30	33.12	37.80	2,300	kWh/day	≤	3,000
Total cost (Euros) 1,362,000							
Number of dwellings to be built by hectare	34	30	10	74	Total number of houses		

Table 8.11 Construction cost per m² of floor space

	Type 1 house	Type 2 house	Type 3 house
Construction cost	16,800	22,000	26,500
Minimum floor space:	47	52	80
Euros/m² of floor space:	357.44	**423.10**	331.25

105,000 l/day (boxed cell) (Table 8.12), instead of 115,000. The entrepreneur knows that said reduction would affect people in his development, and requests information about how badly that problem will influence his business. Modifying the decision matrix introducing the new value in row '*Water consumption*', and running the model again, the Solver displays this message *"Solver has not found a feasible solution"*.

Why?

Examining the data in the decision matrix, the reason for this message immediately appears, because, with that volume of water, the criterion *"Minimum number of Type 2 houses"* is not satisfied, since the minimum threshold is now 31 houses, which is below the minimum level established of 35 houses (boxed cells), and because of that it violates the restriction imposed by the action.

8.3.3.3 Comments on This Case

This case addresses a particular scenario, which is housing development. As that, it introduces restrictions with both lower and upper thresholds, and in fact reflecting actual situations where uncertainties exist about what will be peoples' choices for buying a house. This example also illustrates how to apply a sensitivity analysis to different issues. Another important feature is the determination of which are the most significant criteria, knowledge that could be an invaluable tool for the DM; this information can be obtained using the dual property of the method. It is important to point out that sensitivity analysis can give the DM a lot of information with which he/she can elaborate Costs/Benefits analysis to find the most convenient strategy.

8.3.4 Case Study: Municipal Projects

8.3.4.1 Background Information – General Objective

A city has assigned a certain amount of money to execute civil works in three different areas, with ten different projects. Areas are ***Stormwater, Green Spaces*** and ***Spatial Planning***. Each area has its own budget that has to be respected, and within this budget there are also maximum funds for the different projects.

Table 8.12 Data reducing availability of water supply from 115,000 m³ to 105,000 m³

Urban development

	Plan for type 1 house	Plan for type 2 house	Plan for type 3 house	Result from computation	Units	Action	Requirements
Investment per dwelling (Euros):	16,800	22,000	26,500				
Area of each lot (m^2)	69	122	160	7,402	m^2	≤	7,700
Minimum number of units of type 1 houses	1			30	dwellings	≥	30
Maximum number of units of type 1 houses	1			30	dwellings	≤	40
Minimum number of units of type 2 houses		1		31	dwellings	≥	35
Maximum number of units of type 2 houses		1		31	dwellings	≤	48
Minimum number of units of type 3 houses			1	10	dwellings	≥	5
Maximum number of units of type 3 houses			1	10	dwellings	≤	10
Minimum number of dwellings	1	1	1	71	dwellings	≥	50
Maximum number of dwellings	1	1	1	71	dwellings	≤	95
Maximum total investment (Euros/ha)	1	1	1	1,442,149	Euros/ha	≤	1,500,000
Minimum total investment (Euros/ha)	1	1	1	1,442,149	Euros/ha	≥	1,250,000
Maximum floor space area (m^2/house)	55	65	95	4,588	m^2	≤	5,000
Minimum floor space (m^2/house)	47	52	80	3,800	m^2	≥	3,800
Maximum density (persons/house)	4.55	5.52	6.30	368	persons	≤	400
Minimum density (persons/house)	4.00	5.20	6.15	341	persons	≥	340
Water consumption (liters/day-house)	1,297	1,573	1,796	105,000	liters/day	≤	105,000
Sewage production baths & kitchens (m^3/day-house)	0.98	1.19	1.36	80	m^3/day	≤	90
Electric energy consumption (kWh/day-house)	27.30	33.12	37.80	2,211	kWh/day	≤	3,000
Total cost (Euros)							
Number of dwellings to be built by hectare					Total number of houses		

Message from the model: *'Solver has not found a feasible solution'*

8.3.4.2 Affected Areas and Corresponding Projects

Area: Stormwater

Comprises three projects intended to eliminate flooding in a part of the city during heavy rains, which are:

Projects I.D.	Projects
A	NW-SE stormwater collector,
B	N–S stormwater collector,
C	Construction of an underground water storage facility, to be used as a temporary reservoir to hold the water volume generated in short time during heavy rain, in order to avoid overloading the main trunks and subsequent potential flooding.

Area: Green Spaces

The city has a low ratio of green space/inhabitant; to increase it two schemes and three projects are planned, as follows:

Projects I.D.	Projects
D	Construction of a linear park in stage one,
E	Expansion of the linear park in stage two,
F	Creating a botanical garden.

Area: Spatial Planning

Plans include the following four projects:

Projects I.D.	Projects
G	Construction of an industrial park,
H	Construction of a daycare centre,
I	Construction of dwellings for low income people,
J	Construction of a neighbourhood community centre.

There are 24 restrictions or criteria to which these 10 projects have to comply.

8.3.4.3 Decision Matrix and Solution

The objective is to make the best appropriation of funds in order to minimize costs. Table 8.13 shows the Excel spreadsheet on which data is loaded. As usual with LP, projects are in columns while criteria are in rows. Note the following:

- Cost for each project is indicated in *'Project cost'* row as well as units in next row; the '1 s' scores indicate relationships between a criterion and alternatives. For instance in criterion '*Maximum length of NW-SE stormwater collector*' the '1'

Table 8.13 Data for municipal projects and selection of projects to be executed when funds availability in the timeframe match funds needed

Municipal Projects

AREAS	STORMWATER			GREEN SPACES			SPATIAL PLANNING							
PROJECTS I.D.	A	B	C	D	E	F	G	H	I	J				
Project cost	3,589	3,892	2,654,789	2,259	2,321	108,741	4,593	456,423	22,400	375				
Cost unit	Euros/m	Euros/m	Euros	Euros/ha	Euros/ha	Euros/ha	Euros/ha	Euros	Euros/dwelling	Euros/m²				
Restrictions, conditions or criteria to be met by projects											Results from computation	Action	Requirements	
Stormwater collector														
Maximum budget for NW-SE stormwater collector	1										4,306,800	=	4,306,800	[Euros]
Maximum budget for N-S stormwater collector		1									3,600,100	=	3,600,100	[Euros]
Maximum length of NW-SE stormwater collector	1										1,200	≤	1,200	[meters]
Maximum length of N-S stormwater collector		1									925	≤	925	[meters]
Underground temporary stormwater storage tank			1								2,654,789	=	2,654,789	[Euros]
Maximum budget for stormwater projects	1	1	1								10,561,689	=	10,561,689	[Euros]
Linear park														
Max. available land in 1st. stage				1							10	≤	10	[hectares]
Max. available land in 2nd. stage					1						25	≤	25	[hectares]
Maximum available land in both stages				1	1						35	≤	35	[hectares]
Total budget for linear park				1	1						80,615	=	80,615	[Euros]
City botanic garden														
Minimum technical size of plot						1					6.5	≥	4	[hectares]
Maximum land available						1					6.5	≤	6.5	[hectares]
Maximum funds for botanic garden						1					706,817	≥	706,817	[Euros]

	A	B	C	D	E	F	G	H	I	J				
Industrial park														
Minimum economic size of area to be developed							1				89	≥	60	[hectares]
Maximum land available for industrial park							1				89	≤	89	[hectares]
Daycare centre														
Minimum number of daily patients to be served								250			250	≥	250	[patients/day]
Maximum number of daily patients to be served								420			420	≤	420	[patients/day]
Low -income dwellings														
Minimum economic number of houses in plot									1		289	≥	156	[dwellings]
Maximum physical number of houses in plot									1		289	≤	289	[dwellings]
Total funds for spatial planning							1	1	1		6,922,160	≤	7,338,800	[Euros]
Community centre														
Max. floor area (according to uses and population)										1	450	≤	450	[m^2]
Min. floor area (according to economic reasons)										1	450	≥	225	[m^2]
Total funds for community center										1	168,750	≤	168,750	[Euros]
Total availability of funds for all projects	1	1	1	1	1	1	1	1	1	1	18,856,671	=	**18,856,671**	[Euros]

R E S U L T S

(Headings from above have been replicated only to facilitate reading)

AREAS	STORMWATER			GREEN SPACES			SPATIAL PLANNING			
PROJECTS	A	B	C	D	E	F	G	H	I	J
FINAL RESULT	1,200	925	1	10	25	6.5	89	1	289	450
	meters	meters	number	hectares	hectares	hectares	hectares	number	dwellings	m^2

Total minimum cost (Euros): 18,856,671

establishes the relationship between this criterion and the *'NW-SE stormwater collector'*. In row *'Maximum budget for stormwater projects'*, there are three '1 s' to indicate that the three projects pertaining to 'Stormwater Area' have to be considered in this criterion; in other words, it informs the model that the total available amount of 10.561.689 Euros (See the *'Requirements'* column), must be shared between the three projects for the stormwater collectors.

- The model contemplates that the construction of the two stormwater collectors must precede the construction of the reservoir, for obvious reasons. This restriction has been incorporated in row *'Final result'* (the row vector for solution), by placing A>C and B>C in the Solver window.
- *(Please consider that in general these are not the letters to be placed in the Solver window, but those which correspond to them in the Excel spreadsheet coordinates).*
- In turn, the model needs to be told that stage 2 of Linear Park cannot be chosen if stage 1 is not selected. This is also indicated in row *'Final result'*, by placing in the Solver window: D>E.

8.3.4.4 Comments on This Case

Two circumstances can happen in considering this problem:

A. The first one corresponds to a situation in which funds needed are matched by funds availability; that is there are funds available for the whole series of projects. Refer to Table 8.13.

Note that the total investment (adding up the total values for: Stormwater collectors, Linear park, Botanic garden, Industrial Park, Day care centre, Low-income dwellings and Community centre) amounts to 18,856,671 Euros ,shown in the '*Requirements*' column. Naturally, in this case, because the matching between funds needed and available, the total amount computed by Solver corresponds to the same amount in column '*Results from computation*'. Thus, '*Requirements*' column logs the *thresholds* values, while '*Results from computation*' column logs the *outcome* of the computation; however, it could have not been this way. Suppose that there were criteria whose thresholds indicate this same amount, but discriminated by years, according to cost computations and corresponding schedule for the implementation for each plan, and establishing percent completion as a function of time, assuming for instance that the total plan will take 3 years to materialize.

In this circumstance, it could very well be that the estimated percentage complete, say for year 2011 for the stormwater collector, does not match funds available for that year, that is, there is no agreement between the availability of funds during 2011 with the demands for funds produced by the scheduling of stormwater collectors and other undertakings.

In this circumstance, very possibly the model will indicate that there is no solution, precisely because such concordance does not exist. If this happens, the DM can make adjustments to try to get this concordance – if possible – and before starting the project, possibly by adjusting the schedule.

In this case it also could be feasible to change the '=' signs to the '≤' signs, allowing the model to consider amounts of work less than the available funds for that period. The ideal is of course that values in both above-mentioned columns coincide, but this rarely happens.

B. The second circumstance relates with the situation when there are not enough funds for the series of projects (Table 8.14).

Assume now another typical situation, in which available funds are in fact less than needed, because for instance a financial contribution that was supposed to come from the central government will not materialize, or because City Hall estimates that collection from taxes will be less than expected. Suppose that the total amount available is in reality only about 84% of that needed, that is 15,789,412 Euros. In this case, place this amount in the boxed cell in '*Requirements*' column and run the Solver. The result will again be shown in the '*Final result*' row.

The result has changed because of funds reduction, and now a project like '*Construction of dwellings for low-income people*' (I) takes its minimum value (156 houses), instead of the maximum (289 houses) as in the original scenario. The same happens with project '*Construction of community centre*' (J), which now has dwindled to 270 m^2 instead of 450 m^2 as originally intended. Note also, that project '*Construction of industrial park*' (G), is assigned now 60 ha instead of 89 ha as in the original plan with full financing.

It can be understood that what the model did was to arrange the funding considering the cost minimization objective. The total amount assigned is 15,676,654 Euros in the corresponding dash-boxed cell in '*Results from computation*' column. Therefore, the model used more than 99% of available funds.

What if one of the projects was an electoral campaign promise that has to be honored, either selected or not by the model? Say for instance that said project was '*Construction of daycare center*'. In this case since this project has to be selected no matter what the model suggests, it is necessary to inform the model about this condition. To do this, place in Solver window H = 1, and when running the Solver this project will be always selected together with others.

8.3.4.5 Sensitivity Analysis

There are many areas where sensitivity analysis can be applied, and perhaps one of the most frequent is in financing. Normally municipal projects obey a Master Plan involving long periods of time, with politics and politicians involved, that is, projects envisaged by one administration, could be considered irrelevant and superfluous by the following one and cancelled, and this fact demands a reformulation of the Master Plan with new projects, different funding and different interests. Again, it is considered that in this type of projects the fundamental issues are related to funding, and most of the time at the moment of developing a Master Plan there are doubts, uncertainties and various expectations regarding this issue, which often calls for the development of alternative schemes.

Table 8.14 Data for municipal projects and selection of projects to be executed when funds availability in the timeframe does not match funds needed

	Municipal Projects													
AREAS	STORMWATER			GREEN SPACES			SPATIAL PLANNING							
PROJECTS	A	B	C	D	E	F	G	H	I	J				
Project cost	3,589	3,892	2,654,789	2,259	2,321	108,741	4,593	456,423	22,400	375				
Cost unit	Euros/m	Euros/m	[Euros]	Euros/ha	Euros/ha	[Euros/ha]	Euros/ha	[Euros]	[Euros/dwel.]	[Euros/m^2]				
Restrictions, conditions or criteria to be met by projects											Results from computation	Action	Requirements	
Stormwater collector														
Maximum budget for NW-SE stormwater collector	1										4,306,800	=	4,306,800	[Euros]
Maximum budget for N-S stormwater collector		1									3,600,100	=	3,600,100	[Euros]
Maximum length of NW-SE stormwater collector	1										1,200	≤	1,200	[metres]
Maximum length of N-S stormwater collector		1									925	≤	925	[metres]
Underground temporary stormwater storage tank			1								2,654,789	=	2,654,789	[Euros]
Maximum budget for stormwater projects	1	1	1								10,561,689	=	10,561,689	[Euros]
Linear park														
Max. available land in 1st. stage				1							10	≤	10	[hectares]
Max. available land in 2nd. stage					1						25	≤	25	[hectares]
Maximum available land in both stages				1	1						35	≤	35	[hectares]
Total budget for linear park				1	1						80,615	=	80,615	[Euros]
City botanic garden														
Minimum technical size of plot						1					6.5	≥	4	[hectares]
Maximum land available						1					6.5	≤	6.5	[hectares]
Maximum funds for botanic garden						1					706,816	≥	706,817	[Euros]

	A	B	C	D	E	F	G	H	I	J				
Industrial park														
Minimum economic size of area to be developed							1				60.0	≥	60	[hectares]
Maximum land available for industrial park							1				60.0	≤	89	[hectares]
Daycare centre														
Minimum number of daily patients to be served								250			250.0	≥	250	[patients/day]
Maximum number of daily patients to be served								420			420.0	≤	420	[patients/day]
Low -income dwellings														
Minimum economic number of houses in plot									1		156.0	≥	156	[dwelings]
Maximum physical number of houses in plot									1		156.0	≤	289	[dwelings]
Total funds for spatial planning							1	1	1		6,788,963	≤	7,338,800	[Euros]
Community centre														
Max. floor area (according to uses and population)										1	270	≤	450	[m^2]
Min. floor area (according to economic reasons)										1	270	≥	225	[m^2]
Total funds for community center										1	101,130	≤	168,750	[Euros]
Total availability of funds for all projects	1	1	1	1	1	1	1	1	1	1	15,676,654	=	15,789,412	[Euros]

RESULTS

(Headings from above have been replicated only to facilitate reading)

AREAS	**STORMWATER**			**GREEN SPACES**			**SPATIAL PLANNING**			
PROJECTS	A	B	C	D	E	F	G	H	I	J
FINAL RESULT	1200	925	1	10	25	6.500	60	1	156	270
	[metres]	[metres]	[number]	[hectares]	[hectares]	[hectares]	[hectares]	[number]	[dwellings]	[m^2]

Total minimum cost (Euros): 15,676,654

The case exemplified here precisely pertains to this category, since the amount of funding initially considered was further reduced, and even during the Master Plant preparation stage. Here the model is useful in determining which projects should be considered when there is shortage of funds.

8.3.5 Case Study: Scheduling Bridge Repairs

8.3.5.1 Background Information – General Objective

This actual case corresponds to a rural area in Eastern Canada. The capital of the district is a small town (A), no more than 15.000 inhabitants, acting as the central place[4] of a region with six hamlets (B, C, D, E, F, and G), which spatial location is sketched in Fig. 8.3.

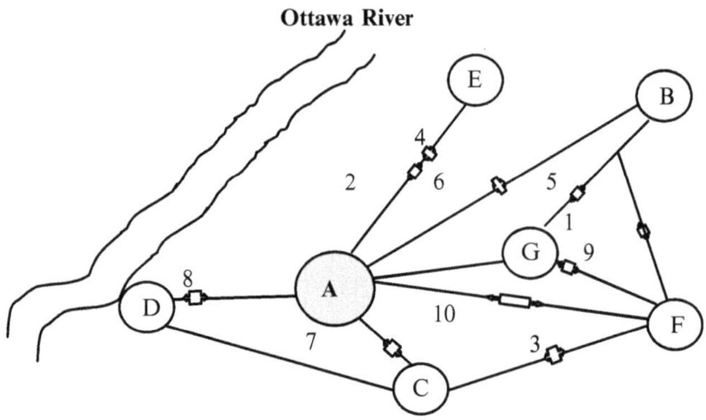

Fig. 8.3 Scheme of the analyzed region

The region is very well connected by good roads and crossed by a series of small rivers and creeks discharging into the Ottawa River. This particular hydrography made imperative the construction of ten bridges of various sizes in order to connect the diverse hamlets. Structures encompassed different types of bridges (steel, wooden, concrete) built along several periods.

There are some old bridges and others newer, and all of them with the capacity of supporting various loads and in diverse states of repair (consider that the oldest bridges had a different design and were built under different demands and load patterns, unlike those built later, and without comparison with the present day heavy traffic and its loads).

[4]From the *'Central Place Theory'*, developed by German geographer Walter Christaller, who tried to explain the location of human settlements and markets. The central place in this theory is a city that provides services to the rest of the area and is usually identified as the most populated community, which normally is the location for regional offices as well as an administrative centre.

8.3.5.2 Repair Schemes

The City Hall in A, responsible for the whole region, has established a plan to repair all bridges and eventually replace the oldest. There is a series of potential different periods or spans for each bridge, identified as follows:

Repair Periods

- Immediate, which in turn involves:
  - (H) High priority,
  - (M) Medium priority,
  - (L) Low priority.
- In the 1–5 years span,
- In the 6–10 years span.

8.3.5.3 Technical Aspects

A *technical inspection* of each bridge evaluates repairs needed, and from here, a *tentative schedule for repairs* is derived. This appraisal depends on some parameters like state of repair, loads that the structure can support (perhaps even decreasing the load allowance, depending on the timing of the repairs), availability of engineering studies, etc. *Cost calculation* follows, and finally a *decision matrix* prepared using '1 s', in correspondence with the tentative repairs scheduled to indicate potential spans. Fig. 8.4 sketches this procedure, while Table 8.15 is the decision matrix.

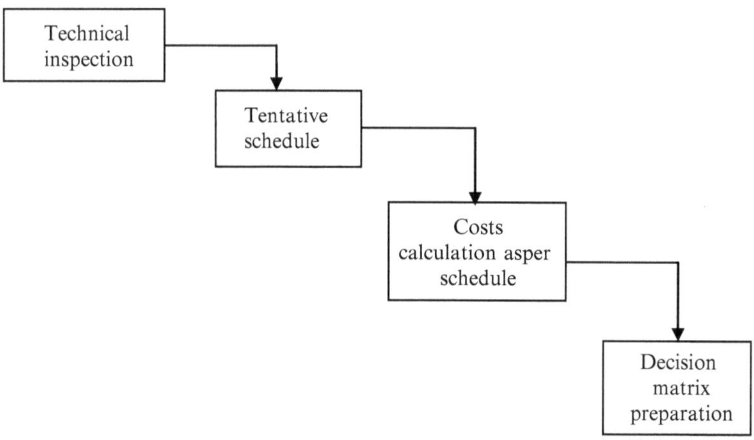

Fig. 8.4 Operative sequence to build the decision matrix

Table 8.15 Data, decision matrix and result for scheduling bridges repair

SCHEDULING BRIDGES REPAIR

DATA

Bridge #:	1	2	3	4	5	6	7	8	9	10		
Costs (CND):	160,000	1,000	194,500	219,000	13,500	164,000	5,000	21,000	120,000	23,500		
People's Priority:	0.50	6	9	1	8	10	7	1	0.50	4		
	1	2	3	4	5	6	7	8	9	10		Sum of rows
Replacement costs (existing)	289,000	538,000	351,000	408,000	206,000	297,000	289,000	336,000	210,000	252,000	=	3,176,000
Replacement costs (Geometric standards)	510,000	538,000	663,000	725,000	293,000	525,000	289,000	336,000	351,000	324,000	=	4,554,000
Repair Costs (H) Immediate	145,000				10,000		5,000		105,000	15,000	=	280,000
Repair Costs (M) Immediate						149,000					=	149,000
Repair Costs (L) Immediate		1,000	176,000		1,000					5,000	=	183,000
Repair Costs at 1-5 span		5,000			15,000		25,000	21,000		1,000	=	67,000
Repair Costs at 6-10 span			30,000		20,000			10,000			=	60,000
Eng. Studies costs (H) Immediate	15,000				2,500				15,000	3,500	=	36,000
Eng. Studies costs (M) Immediate				15,000		15,000					=	30,000
Eng. Studies Costs (L) Immediate			18,500								=	18,500
Eng. Studies Costs at 1-5 years span		5,000			5,000		5,000	7,500		5,000		
Current Posted Load	3		5	5		15			9			
Suggested Posted Load	2		5	3		3			2			

DECISION CRITERIA

FINANCIAL CRITERION: BOXES INDICATE IF PROVINCIAL AND FEDERAL CONTRIBUTIONS ARE AVAILABLE

Bridge #:	1	2	3	4	5	6	7	8	9	10	Sun of rows
Repair and engineering costs:	160,000	1,000	194,500		13,500	164,000	5,000	0	120,000	23,500	681,500
Federal and provincial contribution	106,560	666	129,537	New	8,991	109,224	3,330	0	79,920	15,651	
Cost less fed. & prov. contribution	53,440	334	64,963	Bridge	4,509	54,776	1,670	0	40,080	7,849	2,108,561

REPLACEMENT COST **CRITERION:**		**BOXES INDICATE REPLACEMENT WITH GEOM. COST IF THIS IS LESS THAN 1.5 TIMES REPLACEMENT COSTS (EXISTING)**											
												Data	
Replacement Costs	**289,000**	**538,000**	**351,000**	**New**	**206,000**	**297,000**	**289,000**	**336,000**	**210,000**	**252,000**	≤	**2,768,000**	
Replacement Costs	**289,000**	**538,000**	**351,000**	**bridge**	**293,000**	**297,000**	**289,000**	**336,000**	**210,000**	**324,000**	≥	**2,927,000**	
Decision taken	**0**	**0**	**0**		**1**	**0**	**0**	**0**	**0**	**0**			
ECONOMIC **CRITERION:**													
Total costs with: replacement, federal and province contributions, repairs, and engineering studies	**53,440**	**334**	**64,963**		**137,196**	**54,776**	**1,670**	**0**	**40,080**	**7,849**	≤	**Sum of row 360,308**	
REPAIRS COSTS **CRITERION:**		**BOXES INDICATE**	**IMMEDIATE REPAIR + ENG. STUD.] OR [REPAIR COSTS IN 1-5 AND 6-10 YEARS + ENG. STUD. IN 1-5YEAR**										
Repairs	**160,000**	**1,000**	**194,500**	**New**	**13,500**	**164,000**	**5,000**	**0**	**120,000**	**23,500**			
Repairs	**160,000**	**36,000**	**194,500**	**Bridge**	**13,500**	**164,000**	**5,000**	**0**	**120,000**	**23,500**			
LOAD **CRITERION:**		**BOXES INDICATE REPAIRS NOT DONE NOW, THEN HINTED POSTED LOAD; IF NO REPAIRS NOW AND CURR. POSTED LOAD**											
Load Posted	**3**		**5**	**New**		**15**			**9**		≤	**32**	**years**
Load Posted	**3**		**5**	**bridge**		**15**			**9**		≤		**years**
EXECUTION TIMING **CRITERION:**													
Execution Period in 2011													
June	**0.25**					**0.50**							
July	**0.25**	**0.75**				**0.30**				**0.30**			
August	**0.50**	**0.20**	**0.33**	**New**		**0.20**				**0.20**			
September		**0.05**	**0.33**	**bridge**			**0.10**			**0.10**			
October			**0.34**				**0.20**		**0.10**	**0.10**			
November					**1.00**		**0.20**		**0.30**	**0.05**			
December								**0.35**	**0.20**				
January 2012							**0.50**	**0.65**	**0.40**	**0.25**			
	1	**1**	**1**		**1**	**1**	**1**	**1**	**1**	**1**			

(continued)

Table 8.15 (continued)

DECISION MATRIX AND RESULT

Investments per repair periods:	1,000	10,000	30,000	194,500	--	--	219,000	--	--	13,500	20,000	20,000	164,000	--	--
		Bridge 2			Bridge 3			Bridge 4			Bridge 5			Bridge 6	
Type of policy:	mmediat	1-5 span	6-10 span	Immediate	1-5 span	6-10 span	Immediate	1-5 span	6-10 span	Immediate	1-5 span	6-10 span	mmediat	-5 spa	-10 span
People's Priority:		6			9			1			8			10	
Tentative schedule	1			1						1			1		
Tentative schedule		1						New bridge			1				
Tentative schedule						1						1			
Total budget	1	1	1	1	1	1				1	1	1	1	1	1
Alternatives selection for bridge 2	1	1	1												
Alternatives selection for bridge 3				1	1	1									
Alternatives selection for bridge 4															
Alternatives selection for bridge 5										1	1	1			
Alternatives selection for bridge 6													1	1	1
Alternatives selection for bridge 7															
Alternatives selection for bridge 8															
Alternatives selection for bridge 9															
Alternatives selection for bridge 10															
Bridges:		Bridge 2			Bridge 3			Bridge 4			Bridge 5			Bridge 6	
Type of policy:	Immed.	1-5 span	6-10 span	Immed.	1-5 span	6-10 span	Immed.	1-5 span	6-10 span	Immed.	1-5 span	6-10 span	Immed.	-5 spa	-10 span
SPANS SELECTED BY BRIDGE	0	0	1	1	0	0	0	0	0	0	1	0	1	0	0

(Table continues below)

	5,000	30,000	--	--	28,500	10,000	120,000	--	--	23,500	1,000	-			
		Bridge 7			Bridge 8			Bridge 9			Bridge 10		Results from		Requirements
	Immed.	1-5 span	6-10 span	Immed.	1-5 span	6-10 span	Immed.	1-5 span	6-10 span	Immed.	1-5 span	6-10 span	computation		
		7			1			0.50			4			Operator	
Tentative schedule	1						1			1			721,048	≤	900,500
Tentative schedule		1			1						1		58,225	≤	94,500
Tentative schedule						1							49,863	≤	60,000
Total budget	1	1	1	1	1	1	1	1	1	1	1	1	895,000	≤	1,055,000
Alternatives selection for bridge 2													1	≤	1
Alternatives selection for bridge 3													1	≤	1
Alternatives selection for bridge 4													1	≤	1
Alternatives selection for bridge 5													1	≤	1
Alternatives selection for bridge 6													1	≤	1
Alternatives selection for bridge 7	1	1	1										1	≤	1
Alternatives selection for bridge 8				1	1	1							1	≤	1
Alternatives selection for bridge 9							1	1	1				1	≤	1
Alternatives selection for bridge 10										1	1	1	1	≤	1
Bridges:		Bridge 7			Bridge 8			Bridge 9			Bridge 10				
Type of policy:	Immed.	1-5 span	6-10 span	Immed.	1-5 span	6-10 span	Immed.	1-5 span	6-10 span	Immed.	1-5 span	6-10 span	Total investment:		895,000
SPANS SELECTED BY BRIDGE	0	1	0	0	1	0	1	0	0	1	0	0	Efficiency:		0.85

8.3.5.4 Financing

It comes from two sources: Federal and provincial funds, and in equal amounts per period, therefore, expenditures per month must not exceed these values. Then, it is necessary to associate these disbursements with the monetary valuation corresponding to percentages of completion per month and per bridge (that is, the cash flow), during a period from June to December 2011.

What is the objective?

8.3.5.5 Objective

The objective calls for establishing a schedule of repairs to make the best possible use of limited resources, and subject to construction and financial conditions. The result must be the schedule of repairs for each bridge, i.e. the variables in this case are not the bridges but *periods of repairs (spans), or 'when'*. All bridges are included with the exception of bridge number 1, which will be most probably dismantled (although appraisals for repairs have been made), and number 4, which is brand new.

8.3.5.6 Data, Decision and Solution

Table 8.15 shows the decision table. Also displays (from top to bottom):

Data

- The total available funds amounted to CND 1,055,000 (boxed), in *'Total budget'* row,
- Costs in Canadian dollars (CND), for each bridge,
- People's priority expressed in numbers (after a survey and conversion of subjective values into cardinal),
- Replacement costs as they are now,
- Replacement with costs for geometric[5] standards,
- Immediate repair costs for all bridges that require *high priority* and their summation,
- Immediate repair costs for all bridges that require *medium priority* and their summation,
- Immediate repair costs for all bridges that require *low priority* and their summation,

[5]Costs for geometric standards. Refers to costs related with the geometric design of the whole undertaking, that is including not only the bridge structure but all other elements such as traffic approaching and leaving the bridge, distribution, loads, etc. Naturally geometric costs are higher than just structural costs.

- Costs for all bridges that are scheduled for the 1–5 years span and their summation,
- Costs for all bridges that are scheduled for the 6–10 years span and their summation,
- Engineering studies costs for all bridges that require high, medium and low priority, and their summation,
- Engineering studies costs for 1–5 years span. There is no row to show these costs for the 6–10 years span because it is considered that it is difficult to make an so early an estimate of time.

Decision Criteria

Several types of criteria are considered as follows:

- *Financial criterion*, related with funds availability, i.e. informs about provenance of funds, that is, if they come from the provincial and/or the federal government. If they do, 2/3 of the total amounts are shown here,
- Amounts that must be paid by City Hall,
- *Replacement cost criterion*. For each bridge it makes this analysis: If geometric costs are 1.5 times larger than replacement costs, then adopt replacement costs, otherwise go to geometric costs,
- *Economic criterion* computes for each bridge the cost to be assumed by City Hall once the provincial and federal contributions are considered. Its summation can be then compared with City Hall's availability of its own funds.
- *Repairs cost criterion* lists costs for different scenarios,
- *Load criterion* detailing current posted load for each bridge. It relates selection time for repairs with load posted,
- *Execution timing criterion* related to work to be performed from June 2011 to January 2012 inclusive. Each row details total percentage to be executed in each month, while the summation of any column indicates what percentage of a bridge needs to be done in each month. Why this? To make sure that during a certain month the amount done will not exceed the established percentage which is necessary for cash flow reasons. The summation in columns guaranties that the bridge work is finished.

Decision Matrix and Result

- This decision matrix has been reduced because, in its original format, it was too big for this book's purposes, whose sole objective is to show the methodology. '1 s' correspond to values from data.
- '*Total budget*' row informs us that total investment will be 895,500 CND, which means that 85% of funds have been appropriated. Consequently, the schedule allocates bridge repairs in the most advantageous period, and considering some

requirements. The actual problem solved in the same manner took into consideration all criteria.

- Running Solver, results appear in row *'Spans selected by bridge'*, indicated with '1 s'.
- Observe that the results are expressed in binary values, which are '1' for executing in a certain period and '0' (not shown), for not executing. The reason for this is evident since a repair cannot be made in two different spans, since they are exclusive. This is the way to tell the model to select for each bridge only one of the three alternatives, that is immediate, 1–5 years or 6–10 years.
- Immediately below is the objective function, which is expressed as the summation of the products between values of *'Investments'* and *'Results'* rows. It indicates then the total cost when repairs are allocated in the manner shown in the result. This total value maximizes the objective, which is to use the maximum amount of available capital and at the same time comply with all restrictions imposed.
- Results are summarized in Table 8.16.

Table 8.16 Final schedule of repairs

Bridge number	Schedule of repairs
2	In the 6–10 years span
3	Immediately
5	In the 1–5 years span
6	Immediately
7	In the 1–5 years span
8	In the 1–5 years span
9	Immediately
10	Immediately

8.3.5.7 Sensitivity Analysis

There are different areas to perform sensitivity analysis in this case, such as:

- The tentative schedule can be modified, for instance changing the *'Execution timing criterion'* by altering the execution period, that is, repairs can be considered for another time than originally specified, since normally this is not a case of extreme urgency.
- Percentages of completion for different repairs can be varied, changing values in this tentative schedule.
- Financial terms can vary in both amounts and timing.
- Conditions to apply geometric costs can be perhaps modified.

8.3.5.8 Comments on This Case

As we can see, this is a very difficult decision-making problem with 24 options (eight bridges with three options each), and subject to a series of very tough restrictions.

This case has been detailed in certain depth – on purpose – to illustrate how real life situations can be put many times into a mathematical format for its resolution, and also to show how powerful LP is. Of course, it would be naïve to think that the mathematical model in this case or in others represents or is able to represent faithfully real life situations – they are only approximations – and must be taken as such. It can also be appreciated in the examples throughout this book that in general most cases are formulated more or less in a standard format (that is the decision matrix), however, each one is in general different and consequently its set-up is distinct, but the solving process is exactly the same in all cases.

8.3.6 *Case Study: Land Use and Rehabilitation of Abandoned Land*

8.3.6.1 Background Information – General Objective

In Sect. 2.3 comments were made about urban rehabilitation, and this example presents a case, as follows:

An American harbor city suffered as many others the effect of the new wave in sea transport when container shipping started in the middle of the twentieth century. It was also remarked that about the same time the decline of long distance rail passenger service commenced because of air competition. This city had to tackle the problem of deciding what to do with 42 ha of wharves, warehouses and the main railway station when these services were cancelled (another train station was built in a more adequate location, since the old one, with its large array of rails, platforms and warehouses was no longer needed).

8.3.6.2 Data, Decision Matrix and Solution

An international competition was called to develop the Master Plan whose general guidelines contemplated seven different schemes, with constructions as follows:

Scheme A
Corporate Towers – Hotels – Marina- Little Park.
Scheme B
Corporate Towers – High-Rise Housing – Commercial Area in the old railway station.
Scheme C
Government Complex Centre to house all provincial government offices – Large Mall using the old railway station- Small Park.
Scheme D
High-Tech (Biotechnology, Informatics, etc.) – Park and Recreational Area.

Scheme E
High-Tech (Biotechnology, Informatics, etc.) – Convention Centre.
Scheme F
High-Tech (Biotechnology, Informatics, etc.) – University Main Campus.
Scheme G
World Trade Centre – Large Park and Leisure Area.

Schemes are subject to 15 criteria, as indicated in Table 8.17.

Objective: ***Maximize an envelope of benefits for the city***. Scores for the objective function, benefits, advantages and disadvantages of each scheme will be evaluated and weighted (if possible) to obtain an average value.

Table 8.17 Criteria to evaluate schemes

Criteria I.D.	Area	Explanation	Action
1	Transportation	Refers to transportation to and from the city centre.	Maximize connectivity
2	Job generation	Refers to jobs that will be generated during design, construction and operation. There is an estimate of the number of construction opportunities	Maximize
3	Environmental impact	– Aspects such as present day sea view for people whose panoramic vista will be curtailed by these constructions. – Road congestion because increasing future human density in the area.	Minimize
4	Financial feasibility	The idea is for the City Hall to advance money for this undertaking but also to recover it in a reasonable time span, by selling the cleared land	Maximize
5	Aesthetics	The site is close to the colonial and historic area of the city and a gradual blending between the new and the old is sought, albeit each one with its own characteristics.	Maximize
6	Soil permeability	City Hall is conscious that because of the large civil structures there will be a decrease of rain water permeating the soil that can affect aquifers. It is expected that proposals will offer technical solutions to this problem because it is specifically requested in the Bid Documents: Terms and Conditions.	Maximize
7	Water demand	The development will make a dent in the city water treatment plant; therefore it is necessary to minimize this demand. Proposals shall offer ways to collect and purify rain water to be used as potable water.	Minimize

(continued)

Table 8.17 (continued)

Criteria I.D.	Area	Explanation	Action
8	Energy demand	It is believed that when complete the area will house about 50,000 people, who – because of the high prevalent income level – will surely consume more than the average energy, therefore the proposal must consider using solar energy as much as possible.	Minimize
9	Sewage	The city is in no condition to absorb at this moment effluents from this new area. Proposals must present plans to recycle water and treatments for sludge in each large complex.	Minimize production
10	Municipal Infrastructure	It is for the city to build streets and sidewalks. Proposals must offer layouts for streets and avenues wide enough to guaranty easy transit as per city guidelines.	Minimize
11	Links to subway network	Schemes must be as close as possible to two existing subway stations in the periphery of the area.	Maximize
12	Green space recovery	Proposals must consider layout as to provide as much green space as possible. Roof gardens are permitted and encouraged.	Maximize
13	Business activity	Even though there is no intention to develop an isolated area from the downtown area, provision must be taken to assure that at least 10% of land space is devoted to commerce.	Maximize
14	Citizens' opinion	Schemes must also consider citizens' opinion, when results from a survey conducted by City Hall are known.	Maximize
15	City Council opinion	Schemes will be also subject to City Council opinion, therefore each scheme has to take into consideration Council's decisions.	Maximize

Table 8.18 shows the decision matrix. Notice that there are some shaded cells; its purpose is to point out for each row the maximum or the minimum value according to what is requested by the *'Action'* column. What is interesting to observe here is that maximum and minimum values are approximately evenly distributed between the different schemes, which would make it impossible or at least very difficult, to determine which scheme is better than the others.

The model was first run using the envelope benefits for the city as the objective function. The PM and her team had a hard time to figure out the scores for functional variables, that is the different schemes, and one of the reasons is that each

Table 8.18 Data and decision matrix – The same importance or weight is given to each scheme

	Land use and rehabilitation of abandoned land									
Schemes I.D.	**A**	**B**	**C**	**D**	**E**	**F**	**G**			
Estimated benefits envelope for the city	1	1	1	1	1	1	1			
Selection criteria								**Action**	**Results from computation**	**Operator**	**Requirements**
Transportation	0.153	0.185	0.153	0.183	0.207	0.183	0.190	MAX	0.207	≤	0.207
Job generation	0.196	0.218	0.250	0.201	0.164	0.156	0.200	MAX	0.164	≤	0.250
Environmental impact	0.098	0.113	0.079	0.094	0.065	0.113	0.088	MIN	0.065	≥	0.065
Financial feasibility	0.356	0.251	0.401	0.207	0.219	0.190	0.175	MAX	0.219	≤	0.401
Aesthetics	0.185	0.185	0.185	0.190	0.182	0.185	0.200	MAX	0.182	≤	0.200
Soil permeability	0.423	0.025	0.164	0.174	0.423	0.164	0.418	MAX	0.423	≤	0.423
Water demand	0.196	0.196	0.211	0.196	0.256	0.190	0.185	MIN	0.256	≥	0.185
Energy demand	0.251	0.120	0.240	0.230	0.225	0.319	0.142	MIN	0.225	≥	0.120
Sewage	0.196	0.196	0.211	0.101	0.102	0.101	0.101	MIN	0.102	≥	0.101
Municipal infrastructure	0.164	0.170	0.164	0.185	0.404	0.194	0.203	MIN	0.404	≥	0.164
Link to subway network	0.338	0.330	0.330	0.341	0.164	0.341	0.167	MAX	0.164	≤	0.341
Green space recovered	0.196	0.110	0.180	0.196	0.429	0.168	0.302	MAX	0.429	≤	0.429
Business activity	0.201	0.162	0.307	0.223	0.101	0.079	0.190	MAX	0.101	≤	0.307
Citizens' opinion	0.199	0.236	0.289	0.222	0.321	0.299	0.351	MAX	0.321	≤	0.351
City Council opinion	0.201	0.22	0.218	0.22	0.23	0.195	0.205	MAX	0.230	≤	0.230
Projects participation	1	1	1	1	1	1	1	EQUAL	1.000	=	1
Result:					1						
Objective function:	1	**The objective function used the estimated benefits envelope and the same weight for all schemes**									

blends a mix of uses that cannot be measured with the same stick. For corporate buildings it was not too difficult to assess the potential revenues considering the prospective users; the same for the convention centre and for high rise housing, but how to fairly measure the money value of the park or the recreation and leisure area, that translates into intangible benefits for the citizens? How is it possible to evaluate the benefits for the city due to the large number of trees – being so close to downtown – and doing a good job in absorbing CO_2 and releasing oxygen? How will quality of life improve because of these undertakings? There are techniques to perform this appraisal, but the PM believed that they were too subjective and biased.

For this reason the PM arrived at the conclusion that said task would be impractical due to unreliability of data, and that the best way to proceed would be to assign the same score or importance to each scheme, especially considering that all of them are similar in nature and promise to benefit the city. The PM was conscious that this decision was not certainly the best procedure because the discriminatory power of the objective function will be near null, but it had the advantage that the selection will purely reflect the influence of criteria. This problem was solved using LP, with scheme 'E' selected as displayed in Table 8.18 with '1'.

8.3.6.3 Analysis of Different Objectives

The lower part of Table 8.19 illustrates this method applied to this case. The following objectives were chosen:

- Transportation
- Job generation
- Environmental impact
- Financial feasibility
- Municipal infrastructure
- Green space recovery
- Business activity
- Citizens' opinion
- City Council opinion

This problem was solved using SIMUS, running Solver nine times, each instance with a different objective, and the result is shown boxed in the last row.

Notice the use of Guidelines 1 and 2 from Simus (See Sect. 6.2), but no normalization is needed in the pay-off matrix because we are working with binary numbers. Consequently, the selection will pertain to the scheme with the largest sum number; in this case, the model shows a tie between schemes 'C' and 'E' since both have the same value of 4.

Another objective/s could be used to break the tie, however, notice that when applying the envelope benefits as explained above, it also favoured scheme 'E', and consequently, it is believed that this is the choice.

Table 8.19 Decision matrix – payoff matrix and solution

	Land use and rehabilitation of abandoned land										
									Results		
Schemes I.D.	**A**	**B**	**C**	**D**	**E**	**F**	**G**		**from**		
								Action	**computation**		**Requirements**
Selection criteria										**Operator**	
Transportation	0.153	0.185	0.153	0.183	0.207	0.183	0.190	MAX	0.207	≤	0.207
Job generation	0.196	0.218	0.250	0.201	0.164	0.156	0.200	MAX	0.164	≤	0.250
Environmental impact	0.098	0.113	0.079	0.094	0.065	0.113	0.088	MIN	0.065	≥	0.065
Financial feasibility	0.356	0.251	0.401	0.207	0.219	0.190	0.175	MAX	0.219	≤	0.401
Aesthetics	0.185	0.185	0.185	0.190	0.182	0.185	0.200	MAX	0.182	≤	0.200
Soil permeability	0.423	0.025	0.164	0.174	0.423	0.164	0.418	MAX	0.423	≤	0.423
Water demand	0.196	0.196	0.211	0.196	0.256	0.190	0.185	MIN	0.256	≥	0.185
Energy demand	0.251	0.120	0.240	0.230	0.225	0.319	0.142	MIN	0.225	≥	0.120
Sewage	0.196	0.196	0.211	0.101	0.102	0.101	0.101	MIN	0.102	≥	0.101
Municipal infrastructure	0.164	0.170	0.164	0.185	0.404	0.194	0.203	MIN	0.404	≥	0.164
Link to subway network	0.338	0.330	0.330	0.341	0.164	0.341	0.167	MAX	0.164	≤	0.341
Green space recovered	0.196	0.110	0.180	0.196	0.429	0.168	0.302	MAX	0.429	≤	0.429
Business activity	0.201	0.162	0.307	0.223	0.101	0.079	0.190	MAX	0.101	≤	0.307
Citizens' opinion	0.199	0.236	0.289	0.222	0.321	0.299	0.351	MAX	0.321	≤	0.351
City Council opinion	0.201	0.22	0.218	0.22	0.23	0.195	0.205	MAX	0.230	≤	0.230
Projects participation	1	1	1	1	1	1	1	EQUAL	1.000	=	1
Result:					1						
								When using *'benefits envelope' select scheme 'E'*			

(continued)

Table 8.19 (continued)

Payoff matrix

Schemes I.D.	A	B	C	D	E	F	G	
Results according to following objectives:								Action
Transportation					1			MAX
Job generation			1					MAX
Environmental impact					1			MIN
Financial feasibility			1					MAX
Municipal infrastructure			1					MIN
Green space recovered					1			MAX
Business activity			1					MAX
Citizens' opinion							1	MAX
City Council opinion					1			MAX
Result:			4		4		1	Also select scheme 'E'

Solution considering several objectives: Schemes 'C' and 'E' have the same values

Objective function 0.321

8.3.6.4 Introducing Congruence

In reality, this problem was a little more complicated, because albeit there were seven different schemes they were not independent, since several structures are shared. Because of this the PM team thought that it could perhaps be advantageous to consider three groups of schemes pairing them as follows:

- Scheme 'A' and scheme 'G' must be developed together, that is the Corporate Tower, Hotels and Marina must be built together with the World Trade Centre and the Large Park and Leisure Area. City Council believes that these schemes complement each other; the 'Little Park' in scheme 'A' will be deleted since scheme 'G' has already a large park and leisure area.
- Schemes 'B' and 'F' are also congruent for it is understood that Corporate Towers, High-Rise Housing and Commercial Area will match the High-Tech, and University Main Campus.
- Schemes 'C' and 'E', that is the Government complex centre and large mall, will complement High-Tech and the Convention Centre. The Small Park in 'C' is deleted.
- Scheme 'D' is independent because it is considered that the High-Tech, Business Centre, and Park leave no room for other developments.

Table 8.20 shows these congruencies by different shading of the pairs of congruent schemes. Congruencies are fed into the model using the Solver main window, analyzing each congruence separately, with the result shown in Table 8.21, which depicts the result for the last congruence analyzed.

8.3.6.5 Sensitivity Analysis

To analyze impact, each congruency is introduced now on top of one another, the Solver run, and result logged. Therefore, to the first congruency is added the second, the Solver run and result logged. Finally the third congruency is added to the other two, the Solver run and result logged.

When considering these accumulated congruencies, Table 8.22 shows results.

As can be seen the sensitivity analysis of this sort, and of course there could be several, allows the PM to study different options and adopt the one which better satisfies the objective/s of the whole project.

8.3.6.6 Comments on This Case

This case corresponds to a large number of rehabilitation projects where the decision is centered on what to do to recuperate declining or degraded areas of a city. Consequently, it refers mainly to urban projects with large tracts of land that can be devoted to different uses, which usually involves a substantial array of different projects, with housing, corporate offices, amenities, malls, etc. One of the difficulties

Table 8.20 Data, decision matrix and results when there is congruence

	Land use and rehabilitation of abandoned land										
Schemes	**A**	**B**	**C**	**D**	**E**	**F**	**G**				
Estimated benefits envelope for the city	1	1	1	1	1	1	1		**Results from**		
Selection criteria								**Action**	**computation**	**Operator**	**Req.**
Transportation	0.153	0.185	0.153	0.183	0.207	0.183	0.190	MAX	0.207	≤	0.207
Job generation	0.196	0.218	0.250	0.201	0.164	0.156	0.200	MAX	0.164	≤	0.250
Environmental impact	0.098	0.113	0.079	0.094	0.065	0.113	0.088	MIN	0.065	≥	0.065
Financial feasibility	0.356	0.251	0.401	0.207	0.219	0.190	0.175	MAX	0.219	≤	0.401
Aesthetics	0.185	0.185	0.185	0.190	0.182	0.185	0.200	MAX	0.182	≤	0.200
Soil permeability	0.423	0.025	0.164	0.174	0.423	0.164	0.418	MAX	0.423	≤	0.423
Water demand	0.196	0.196	0.211	0.196	0.256	0.190	0.185	MIN	0.256	≥	0.185
Energy demand	0.251	0.120	0.240	0.230	0.225	0.319	0.142	MIN	0.225	≥	0.120
Sewage	0.196	0.196	0.211	0.101	0.102	0.101	0.101	MIN	0.102	≥	0.101
Municipal infrastructure	0.164	0.170	0.164	0.185	0.404	0.194	0.203	MIN	0.404	≥	0.164
Link to subway network	0.338	0.330	0.330	0.341	0.164	0.341	0.167	MAX	0.164	≤	0.341
Green space recovered	0.196	0.110	0.180	0.196	0.429	0.168	0.302	MAX	0.429	≤	0.429
Business activity	0.201	0.162	0.307	0.223	0.101	0.079	0.190	MAX	0.101	≤	0.307
Citizens' opinion	0.199	0.236	0.289	0.222	0.321	0.299	0.351	MAX	0.321	≤	0.351
City Council opinion	0.201	0.22	0.218	0.22	0.23	0.195	0.205	MAX	0.230	≤	0.230
Projects participation	1	1	1	1	1	1	1	EQUAL	7.000	=	1
Result:					1						
Objective function:	1										

Table 8.21 Congruencies independently considered

Congruence	Schemes selected
A = G	E
B = F	E
C = E	E

Table 8.22 Accumulated congruencies

Congruence	Schemes selected with one congruence	Schemes selected with two simultaneous congruencies	Schemes selected with three simultaneous congruencies
A = G	E		
B = F	E	F	
C = E	E		F

of this type of projects is to assert the relative importance of each one, because they could refer to different undertakings, and for this reason in this case, it is suggested to work with the same weight for them all, and in doubt, apply the analysis with multiple objectives where the scores are much easier to determine.

It is also introduced in this case, in the sensitivity analysis section, a different approach, which is for instance grouping two or more undertakings in groups instead of developing them alone. This methodology, depending naturally on particular considerations, has the virtue of packing different projects and at the same time getting more complementarity and probably reducing costs, as well as allowing the constitution of transitory joint ventures between interested construction firms, which brings more expertise and reduced costs.

8.3.7 *Case Study: Contractor's Selection for Metallurgical Development*

8.3.7.1 Background Information – General Objective

This project pertains to the construction of a US$ 331 million copper concentration plant with ore extracted from mines located in government owned land, to produce metallic copper using the Froth-Floatation process, and located high in the Andes Range, in South America. The project sponsor was INDELCO (Industria del Cobre), a mixed capital corporation under the Ministry of Industry and Mines, who hired an American firm as a consultant and project manager. This consultant was also in charge of selecting a local main contractor responsible for the execution of the whole project, which includes building construction, equipment installation and road improvement to a nearby harbor. Construction was a difficult task because of the location of the copper mine and lack of necessary road infrastructure, therefore requirements are tough. The objective was then to select, out of a group of shortlisted contractors, the best general contractor for the job, subject to a series of requirements.

After an initial selection of available general firms, five reputable contractors were shortlisted and invited to send their qualifications and credentials for the project according to the Bid Terms and Conditions. Out of that shortlist only one general contractor would be selected, albeit joint ventures between contractors were allowed. The consultant requested referrals from different sources about these five companies, and based on them compiled an evaluation for each one to be used as scores for the objective function *'Referrals from industry'*.

The five shortlisted contractors were (names have been changed):

- ALARCO SA
- ITEA Construcciones SA
- TEAMIC SAIC
- CONSPAC Pacific SA
- GUILLERMO AGST Construcciones SA

8.3.7.2 Selection Criteria

To evaluate contractors, 15 criteria were chosen as depicted in Table 8.23, all with the same weight.

Table 8.23 Criteria to evaluate contractors

Criteria I.D.	Criteria
1	Years in business
2	Number of engineers
3	Number of other specialties (geology, transportation, etc.)
4	Number of workers for this project
5	M^2 built in the last 3 years
6	Expertise in this project area expressed in number of projects
7	Percentage of own equipment for this project
8	Average age of equipment
9	Number of projects finished in the last 15 years
10	Amount in projects value in millions of Euros in the last 15 years
11	Number of projects delayed more than 10% of initial schedule in 15 years
12	Number of times taken to court in the last 15 years for job related issues
13	Number of suits won
14	Amount of working capital for this project in millions of Euros
15	Liability insurance in millions of Euros

Table 8.24 depicts the decision matrix in Excel and also gives the solution selecting *CONSPAC Pacific SA* as the main contractor in the last row of second matrix.

8.3.7.3 Analysis Using Multiobjective (SIMUS)

Because of the importance of the undertaking, it was also decided to analyze the selection using other objectives (9 in total), which were selected amongst the criteria. They are shown in Table 8.25.

Table 8.24 Data, decision matrix, and its solution – solving for several objectives

		Contractors' Selection					
		ALARCO SA	ITEA CONST. SA	TEAMIC SACIC	CONSPAC Pacific SA	G. AGST Construcciones	
Referrals from industry		7.8	7.6	7.5	7.9	7.8	
Criteria	#	Decision matrix					Sum of rows
Years in business	1	43	52	54	74	67	290
Number of engineers	2	45	49	62	76	39	271
Other technical specialties (geology, transportation, etc.)	3	11	8	0	7	12	38
Number of workers in this project	4	310	294	309	310	290	1,513
Square meters built in the last 3 years	5	140,000	201,000	178,000	208,000	99,000	826,000
Expertise in this project area expressed in number of projects	6	9	12	18	9	13	61
Percentage of own equipment for this project	7	87	82	90	100	94	453
Average age of equipment	8	5	8	3	5	4	25
Number of projects finished in the last 15 years	9	41	54	69	56	45	265
mount in project value in millions of Euros in the last 15 years	10	2.8	2.4	1.9	3.9	2.7	13.7
Projects delayed more than 10 % of initial schedule in 15 years	11	14	21	19	28	12	94
Number of times taken to court in last 15 years	12	8	10	19	23	14	74
Number of lawsuits won	13	5	5	15	12	9	46
Amount of working capital for this project in millions of US$	14	2.8	1.9	2.4	2.1	1.9	11.1
Liability insurance in million of US$	15	10	9	12	15	10	56

(continued)

Table 8.24 (continued)

Criteria	#	Normalization					Results from computation: Action		Operator	Required	Shadow price
Years in business	1	0.148	0.179	0.186	0.255	0.231	MAX	0.255	⊠	0.255	1
Number of engineers	2	0.166	0.181	0.229	0.280	0.144	MAX	0.280	⊠	0.280	
Other technical specialties (geology, transportation, etc.)	3	0.289	0.211	0.000	0.184	0.316	MAX	0.184	⊠	0.316	
Number of workers in this project	4	0.205	0.194	0.204	0.205	0.192	MAX	0.205	⊠	0.205	
Square meters built in the last 3 years	5	0.169	0.243	0.215	0.252	0.120	MAX	0.252	⊠	0.252	
Expertise in this project area expressed in number of projects	6	0.148	0.197	0.295	0.148	0.213	MAX	0.148	⊠	0.295	
Percentage of own equipment for this project	7	0.192	0.181	0.199	0.221	0.208	MAX	0.221	⊠	0.221	
Average age of equipment	8	0.200	0.320	0.120	0.200	0.160	MIN	0.200	⊠	0.120	
Number of projects finished in the last 15 years	9	0.155	0.204	0.260	0.211	0.170	MAX	0.211	⊠	0.260	
Amount in project value in millions of Euros in the last 15 years	10	0.204	0.175	0.139	0.285	0.197	MAX	0.285	⊠	0.285	
Projects delayed more than 10 % of initial schedule in 15 years	11	0.149	0.223	0.202	0.298	0.128	MIN	0.298	⊠	0.128	
Number of times taken to court in last 15 years	12	0.108	0.135	0.257	0.311	0.189	MIN	0.311	⊠	0.108	
Number of lawsuits won	13	0.109	0.109	0.326	0.261	0.196	MAX	0.261	⊠	0.326	
Amount of working capital for this project in millions of US$	14	0.252	0.171	0.216	0.189	0.171	MAX	0.189	⊠	0.252	
Liability insurance in million of US$	15	0.179	0.161	0.214	0.268	0.179	MAX	0.268	⊠	0.268	
Projects participation		1	1	1	1	1	EQUAL	1.000	=	1	
Result using *'Referrals from industry'*					1						
Objective function 0.268											

Payoff matrix

Objectives	#	ALARCO SA	ITEA CONST. SA	TEAMIC SACIC	CONSPAC Pacific SA	G. AGST Construcciones	Action
Years in business	1				1		MAX
Square meters built in the last 3 years	5				1		MAX
Expertise in this project area expressed in number of projects	6	1		1			MAX
Percentage of own equipment for this project	7				1		MAX
Average age of equipment	8					1	MIN
Number of projects finished in the last 15 years	9					1	MAX
Amount in project value in millions of Euros in the last 15 years	10				1		MAX
Amount of working capital for this project in millions of US$	14	1		1			MAX
Liability insurance in million of US$	15				1		MAX
Result using several objectives		2		2	5	2	

Table 8.25 Criteria used as objectives

Criteria I.D.	Objectives
1	Years in business
5	M^2 built in the last 3 years
6	Expertise in this project area expressed in number of projects
7	Percentage of own equipment for this project
8	Average age of equipment
9	Number of projects finished in the last 15 years
10	Amount in projects value in millions of Euros in the last 15 years
14	Amount of working capital for this project in millions of Euros
15	Liability insurance in millions of Euros

8.3.7.4 Analyzing Solutions

Observe the following:

1. The winner can be found in the payoff matrix, row *'Result using several objectives'* in Table 8.24, last row of second matrix, and is again *CONSPAC Pacific SA* for a larger margin compared with other competitors.
2. As seen, the result from a multiobjective method is similar to that found for a single objective; but this is not mandatory and certainly not compelling, although this coincidence produces a comforting feeling. Why? Because it is known for sure that with the single objective the solution found is optimal, that is, it cannot be improved, and if in the multiobjective version – considering nothing less than nine objectives – the same result is achieved, there are solid grounds to deem that this second solution, which of course is not optimal, is nonetheless one of the best.
3. Since both results agree it is worth making a comparison. In the single objective there is not too much information; it just says that the best solution corresponds to contractor *CONSPAC Pacific SA*, and it is right because the model was required to produce only binary results. However, the multiobjective version, even when it is also binary shows that there is a considerable advantage when comparing contractor *CONSPAC Pacific SA* with the others, which are also even.
4. Observe that in rows '*Expertise in this project area expressed in number of projects*' (number 6), and '*Amount of working capital for this project in millions of US$*' (number 14), the model chooses contractors *'ALARCO'* and *'TEAMIC'*, which are in a joint venture. Why?

   It is not difficult to see the reason, observing that for both criteria this pair has combined the highest scores respectively.

5. Notice the value in column *'Shadow price'* in the *'Normalization'* data. The '1' value is the shadow price or marginal value, or the dual of the problem (See Sect. 6.5). It is indicating that a unit variation in said criterion (*Years in business*) has the potential of changing the objective value depicted in the objective function, but it is also indicating that that criterion is the most important of them all.

8.3.7.5 Sensitivity Analysis

- The first analysis using all criteria and the objective function *'Referrals from industry'* takes into consideration the evaluation made by the consultant, *however the DM wants to kno*w what happens if all shortlisted firms are given the same weight. Why does she consider this analysis to be important? Because, placing all companies in the same range of qualifications, the model can tell which of them best complies with all restrictions. Consequently, in row *'Referral from industry'* all the contractors are given the same qualification or weight and the model is run again. This result – not shown – indicates that in this circumstance the better choice is *'ITEA'*. Therefore it is in the DM's hands to decide if referrals from industry are or not more important than criteria compliance.
- The DM demands also to know the ranking among the competing companies. To do this, the condition of obtaining binary results is removed from the Solver window, and the model run again. The result – not shown – gives decimal values which, when considering from the largest to the lowest, gave this ranking:

  First: *ALARCO – TEAMIC* joint venture
  Second: *CONSPAC Pacific*
  Third: *ITEA*
  Fourth: *GUILLERMO AGST*

Considering all this data it is clear that *CONSPACPacific SA*:

- Ranks first when referrals from industry are used as objective function,
- Ranks first when analysis of objectives is done, and by a wide margin,
- Ranks second when ranking different contractors with equal weighting.

Consequently, it appears that this contractor is a good candidate to be selected.

8.3.7.6 Comments on This Case

This case illustrates the decision-making process when it is related with selection of contractors, which also applies to suppliers, personal selection, Enterprise Resources Planning for selecting software and partner, etc. Needless to say, there are many variations that can be considered, however, the procedure using Linear Programming is the same. Normally, in this type of problem, which is not so often in others, there is uncertainty in fixing the scores for the subjective function, and for that reason it is interesting and also useful to run the model assuming the same weights for all alternatives. As in this case, this procedure can give extra information that can be used by the DM in taking his/her decision.

8.3.8 Conclusion on This Chapter

Seven projects have been solved and commented on here. The variety of undertakings has been chosen in order to give, if not a complete – which is a quite impossible task – at least a comprehensive scope of Linear Programming to solve complex projects, and what is believed more important, to show the role of the DM and the possibility of analysis what this tool allows. Most examples also point out the importance of the sensitivity analysis, without which a project analysis is practically useless because of the many uncertainties always present in large projects. It is believed that this type of information can greatly help a reader to become aware of the multiple facets that a project can present.

References

Cohon, J. (1978). *Multiobjective programming and planning* (Mathematics in science and engineering, Vol. 140). New York: Academic Press.

Cohon, J. (2003). *Multiobjective programming and planning*. New York: Dover Publications.

Cohon, J., Facet, T., Haan, A., & Marks, D. (1973). *Mathematical programming models and methodological approaches for river basin planning.* (Technical Report, Ralph M). Cambridge: Parsons Lab for Water Resources and Hydrodynamics, M.I.T.

Hansen, B. (1996). *Fuzzy logic and linear programming find optimal solutions for meteorological problems*, term paper for Fuzzy Logic course at Technical University of Nova Scotia, Canada. http://www.webindia123.com/career/studyabroad/Canada/details.asp?uname=Technical+University+Nova+Scotia-Retrieved 03 Feb 2010.

Shannon, C. (1948). Mathematical theory of communication. *The Bell System Technical Journal, 27*, 379–423, 623–656.

Zeleny, M. (1982). *Multiple criteria decision*. New York: McGraw Hill. 1973.

Chapter 9
SWOT Analysis – Risk Analysis – Actual Problems Solved and Methods Used

Abstract This chapter has been developed as a support for the techniques commented upon in different chapters. It includes two examples, the first designed to demonstrate the use of the SWOT (Strengths, Weaknesses, Opportunities and Threats) analysis. It is a simple example in which the intention is to show how to proceed to evaluate impacts. The second case pertains to risk analysis and it is applied to selecting construction alternatives. The final part of the chapter displays the names of 66 projects, and information about each one such as title, area to which it belongs, author/s name/s, sources where the corresponding paper or article can be found, and their electronic addresses. This table contemplates the following methods: AHP and ANP, ELECTRE, PROMETHEE, Cost/Benefit, LP and SIMUS, MAUT, REGIME, SAMI and some others. Its purpose is to support the DM by giving her/him the opportunity to study different approaches to a problem that could be similar.

Keywords SWOT analysis • Risk analysis • AHP • ELECTRE, PROMETHEE, LINEAR PROGRAMMING • SIMUS

9.1 SWOT Analysis Applied to the Decision Making Process

SWOT is an acronym for Strength, Weakness, Opportunities and Threats. In the decision making process, it is useful to critically examine each project or alternative to learn about its strong and weak points, and to choose a criterion measuring these issues. As an example, assume that there are four or more social alternatives to be considered for selection. Each of them can be assigned a score, based on a criterion such as '*improving peoples' working abilities*' through plans, programs, and courses to train people in different trades. It is likely possible to assign quantitative values (e.g. number of interested persons) to each alternative, or qualitative values (e.g. a subjective evaluation of an alternative), regarding this criterion, and

N. Munier, *A Strategy for Using Multicriteria Analysis in Decision-Making: A Guide for Simple and Complex Environmental Projects*, DOI 10.1007/978-94-007-1512-7_9,

then maximizing it. Of course, if there are disadvantages in each alternative, the criterion will be minimized.

SWOT analysis can be used to find the strengths and weakness of each alternative and scores assigned accordingly. When applied to projects SWOT may have the following meanings:

Strength

It tries to determine what are the project's strong points, i.e. what is really positive, valuable, in the project. A case could be a project that calls for the development of something new and innovative, or has qualities that make it unique. An example is the Millau Viaduct in France, an engineering marvel laid at more than 300 m high, which is the tallest road bridge in the world. It brings prestige for French engineering, facilitates communication in spanning the Tarn River valley, shortening travel time on the Paris-Montpellier route, and opens the road for similar undertakings and to sell French expertise (in allure it looks like a replication of the construction of the Eiffel Tower at the end of the nineteenth century).

Weakness

What are the problems that need to be addressed and improved? Where can the project fail and why?

Weakness often manifests itself in vulnerability of the project, for instance a project for a line of food products with not so large a variety. The business is vulnerable because poor sales of the main product cannot be compensated for by sales on other items, and on the other hand, because of the high dependency of the business on only one product it might not be able to absorb commercial and infrastructure expenses.

Strength and weakness are intrinsic to any project.

Opportunities

What are the good and worthy things external to the project that can help it? This refers to prospects the project may have due to causes not directly related to it. For instance an imported product may have very good opportunities to perform well because of expected government regulations to abolish import taxes.

Threats

What are the bad things that can happen to the project?

For instance, the construction of a subway line in a city which is partially financed by the city, but with most of the funds coming from the federal government, which may be reduced if a different political party assumes power in the next elections. A real life example was the 1993 cancellation by the Congress of the USA for continuation of the Superconducting Super Collider (SSC) effort, when it was under construction. Hints of the threat were known during the design stage because of high costs involved, which provoked heated debates.

Both Opportunities and Threats can be then seen as extrinsic to a project.

Tables 9.1 and 9.2 show an example of this analysis made by a bike manufacturing project.

9.1.1 *Case Study: The ABC Bike Manufacturing Company*

The ABC Company manufactures scooters and now they have decided to enter into the bikes market. The company has plans to manufacture three different lines of products.

(a) Conventional bikes,
(b) Electric bikes with hydrogen fuel cells,
(c) Electric bikes.

The company must decide through a selection process which line to develop with the objective of maximizing profits. The three different lines of products are not exclusive; therefore, any of them or a combination of them can be manufactured with shared firm resources such as investment funds, facilities and equipment, manpower, sales and purchasing departments, etc.

To have an idea of the possibilities, potential, and challenges and after a thorough research, a SWOT analysis was performed for each line of products. Results from this investigation and corresponding data are reflected and condensed in a decision matrix as a base for the decision-making process. The following example (Table 9.1), shows a version of the SWOT analysis made for line (c) Electric bikes.

Table 9.1 SWOT Analysis for ebikes (Intrinsic features: Strengths and Weaknesses)

	Strengths	Weakness	Comments
Weight	It will be the lightest electric bike in the market. The bike body frame is made out of carbon fibres which have large strength and rigidity	With a very strong impact the frame can be broken	Because it is so light it offers less resistance to pedalling and eases the work of the electric motor especially riding uphill
Price		The price is the highest in its category	It will be not the cheapest bike in the market but it will have a ratio of required energy/ distance travelled that favours the product
Materials	It is innovative in the use of high strength plastic and titanium alloys in delicate parts		
Battery	This is its main advantage. Its low weight rechargeable battery has an innovative and patented design that offers more net output and with less recharging period (4 h, instead of the conventional models on the market between 4 and 7 h)		This bike offers a similar braking device as in electric automobiles, for it recovers and stores energy when braking and when going downhill

(continued)

Table 9.1 (continued)

	Strengths	Weakness	Comments
Motor	A robust 450 watts brushless electric motor		We offer two batteries per bike, in order to have always one battery completely charged at home
Sensor	Provides information about the pedalling torque and shows the rider how evenly is his/her pedalling		
Qualifications of this product for an effective market penetration	There is a potential market for this product to be used as a daily commuter to work		

Table 9.2 SWOT Analysis for ebikes (Extrinsic features: Opportunities and Threats)

	Opportunities	Threats	Comments
Cities clogged with traffic	Traffic in most European cities is becoming a nightmare and many people are considering changing their commuting vehicles. This will undoubtedly boost the purchases of motorcycles and bikes as trends from past years show		
Tax incentive	Regulations are under study in many European municipalities to substantially reduce the Value Added Tax when purchasing this type of vehicles		
Import		Competency from imported bikes especially from China	At present time there is a heavy import of electric bikes, and some governments are reluctant to apply import restrictions to protect local industry

(continued)

Table 9.2 (continued)

	Opportunities	Threats	Comments
Environment	There is a strong consensus about the necessity of reducing air contamination, and from this point of view many municipalities are drastically extending their bike exclusive lanes. Some cities have already more than 200 Km, which cover the whole city. This is a trend followed in the Americas too		
Energy		Chinese and UK manufacturers are offering hydrogen fuel cell bikes, which will have an impact on conventional ebikes. However, the threat does not seem immediate because the lack of hydrogen refilling stations which certainly will take some years to develop. This in turn will increase the opportunities for our potential line of hydrogen fuel cells (project b)	It is believed that undoubtedly this could be the future, especially when the hydrogen will be stored in a leak-proof container placed inside the hollow tubular components of the bike's body frame
Market for ebikes	The market is expected to increase at an average of 7% annually in the next 5 years	Because of its potential it is believed that many firms will try to enter into this market, perhaps with less expensive products	However, if ABC 's product goes into the market in a short time it will have a definite advantage over the competition

From this analysis ABC's marketing people can reach some conclusions about those external variables such as demand, price people may be willing to pay, trends, etc. These parameters can then be utilized, together with the corresponding values for the other two product lines (a), (b) and (c), as scores for each alternative in different criteria such as maximum funds available for investment, minimum rate of return (IRR), potential total demand, etc.

This is the type of application for which LP is very well endowed and known, for it will also give a solution with indication of the respective amounts to manufacture of each product to maximize the profit objective.

9.2 How to Estimate Risks

9.2.1 Background Information

A risk is something that can negatively change the scope of work, the completion time, the estimated budget, the safety procedures, etc., therefore its detection and evaluation is fundamental to have a reasonable confidence that a project will develop as planned. We cannot avoid risk but we can and we must try to diminish its influence in a project.

There are no projects without risks, for they exist everywhere. There are risks in not finishing a project on time (very common), risk on exceeding the budget (very common), financial, economic, geological, political, risks, you name it. Therefore what is to be done in this circumstance? First, it is necessary to be able to determine what the risks are and on which stage of the project they may impact. Second, once the risks are known, evaluate them, that is calculate their probabilities, and third, take adequate measures to prevent the conditions under which those risks could manifest themselves. This is done by establishing appropriate safeguards, that is adopting measures that will tend to avoid the risk or at least to decrease it. This information can be quantified as scores for each alternative or project.

This is neither an easy task nor a guarantee that, even with safeguards in place, the project will be risk-free, and to add insult to injury, these measures cost money. This section will illustrate with a fictitious but down-to-earth example of how to proceed. Of course, it is only an example – and should be considered as that – merely designed to outline some possible steps. Since it is impossible to assume no risk at all, the DM must choose a value he/she is satisfied with. In this case let us suppose that the DM is willingness to accept risk is no more than 8%.

9.2.2 Case Study: Selecting Construction Alternatives for a Subway Line to Cross a River

A city wants to build its first subway line. The tunnel will be dug using a TBM (Tunnel Boring Machine), a large (largest machines can span perhaps a little less than a football field) and costly machine that is equipped with a 6 m – or whatever the size of the tunnel is – diameter rotating cutting wheel (cutter head), furnished with discs cutters or wheels that bore the tunnel as the cutter head is pushed against the front rock by a hydraulic mechanism. The removed rock material (muck), is discharged into a conveyor belt behind the machine, which transports the material out of the tunnel. To prevent dangerous rock loosening in a freshly cut section of the tunnel, concrete coming at high speed through a hose and a nozzle (shotcrete), is sprayed onto the recently chiseled tunnel's circumferential sides. This is a transitory protection because immediately after, prefabricated rings of dowels made of concrete and adjusting to each other likes the pieces of a puzzle are placed on the tunnel walls.

This example applies to the line crossing beneath a mighty river, which is a critical phase of the whole project and for that reason three alternatives were preselected. The tunnel, whatever the alternatives, will be 223 m long, with its top at 6.5 m below the river bed (except in alternative c). The three alternatives to choose from are:

(a) Boring two parallel tunnels (for the two tracks). Sketch in Fig. 9.1.

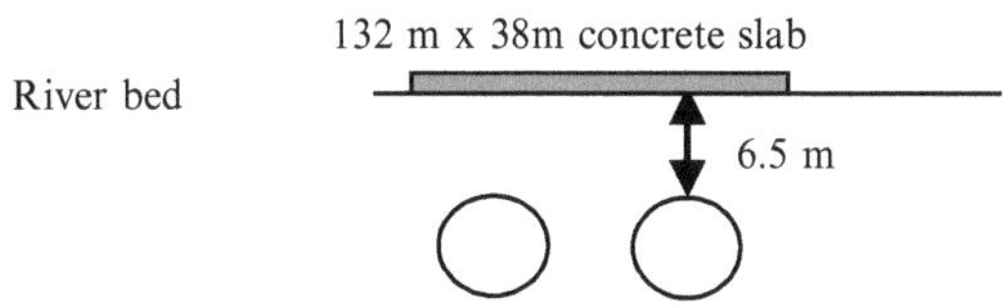

Fig. 9.1 Sectional view of horizontal parallel tunnels

(b) Boring two tunnels one beneath the other. Sketch in Fig. 9.2

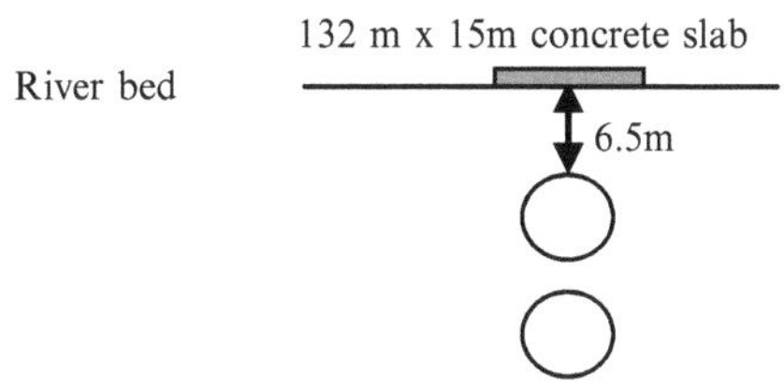

Fig. 9.2 Sectional view of vertical parallel tunnels

(c) Building the tunnel using prefabricated concrete tubes, which are linked and sealed with a concrete collar joint. Sketch in Fig. 9.3.

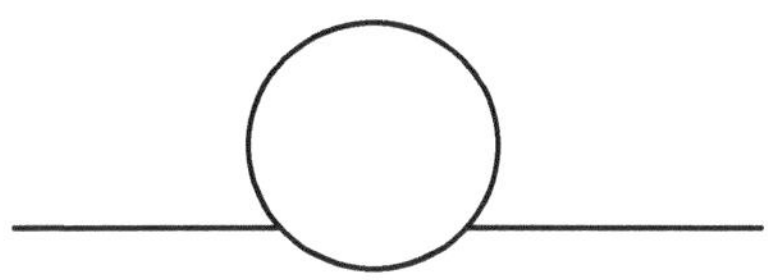

Fig. 9.3 Sectional view of a tunnel tube

In order to analyze these three alternatives, risks are evaluated separately. For this example we are examining only the first alternative (a), of course in a very elementary manner, just to serve as an example of how risks are calculated.

9.2.2.1 Assets Affected

The first action is to determine which assets can be affected if a risk materializes. What are the assets in this project alternative? They are (fictitious numbers, just to work with them). See Table 9.3.

1. The TBM already assembled is worth 5,538,943 Euros,
2. The tunnel itself, which is becoming an asset as its construction advances, with a total cost of 1,586,207 Euros,
3. The concrete plant necessary to produce the dowels costs 341,734 Euros,
4. The cutting discs form the cutter head: 241 Euros/each,
5. The conveyor belt: 56,000 Euros,
6. The electricity plant necessary to guarantee uninterrupted electrical services worth 128,564 Euros.

9.2.2.2 Assets Dependency Value

Considering Table 9.3, the assets have the following accumulated value:

1. TBM: $3,538,943 + 1 \times 56,000 + 1 \times 28,564 = 3,723,507$ Euros
2. Tunnel: $1,586,207 + 1 \times 5,538,943 = 7,125,150$ Euros
3. Concrete plant: $341,734 + 0.5 \times 128,564 = 406,016$ Euros

Table 9.3 Assets and dependency

Assets	Economic value (€)	Depends on:	Reason for dependence	Percent of dependency
1. TBM	5,538,943	5	It also depends on the conveyor belt because the machine needs the extracted material removed as it is generated	100
		6	The TBM is an electricity operated machine; therefore an interruption in the electrical energy stops it	100
2. Tunnel	1,586,207	1	The tunnel will be completely dug by the TBM with a daily cost in all concepts[a] of 86,206 Euros per day, estimating a 30 days' work duration for the 223 m long tunnel	100
3. Concrete plant	341,734	6	Depends on the supply of electrical energy as well as Portland cement and steel (although these are not assets). However the dependence on the electricity plant is not 100% because there is always a stock of dowels for a 10 days operation	50
4. Spare cutting discs	10,845	–	N/A	
5. Conveyor belt	56,000	6	Depends on the supply of electrical energy	100
6. Electric generation plant	128,564	–	N/A	

[a] Such as electric energy, wages, machine amortization, belt conveyor, maintenance, etc.

9.2.2.3 Threats

What are likely to be the foreseen threats here? They are in Table 9.4. Look at the last threat, identified as number 08.

Geological samples indicated that the soil beneath the bed of the river had a poor quality to allow tunnel boring because it consisted of graded aggregate and with a high probability of water percolating, or worse, flooding the tunnel. That was a very serious threat not only for the work, but too for the people working in the tunnel, and who, in case of flooding would not have the opportunity to escape to the surface. For that reason the solution or safeguard found consisted in reinforcing the bed of the river, pouring a thick concrete slab 132 m long by 38 m wide in the case of the two horizontal tunnels, and 15 m wide in case the second option were chosen, as sketched in Figs. 9.1 and 9.2. The execution of this safeguard took 5 months, at a cost of 578,390 Euros.

9.2.2.4 Vulnerability and Degradation – Impact – Intrinsic Risk

Vulnerability: It means how vulnerable or sensitive the asset to a threat is. Its estimate comes from experience, similar works, expert opinions, etc.

Degradation: It means how strong the damage is if vulnerability is broken. Its evaluation comes by appraising asset losses in terms of performing capacity. Table 9.4 shows these values for the Threat/Asset pair, while Table 9.5 shows the values of the Threat/Safeguard pair.

Impact: It is the threat to an asset; computed multiplying the economic value of the asset by the percentage of degradation.

Intrinsic Risk: is the risk of doing nothing to prevent or decrease the impact of detrimental event. It is equal to the product of the impact and the vulnerability. This is a very important concept because it will be used as a reference when safeguards are in place.

Table 9.4 Threats and assets pairs

Threat/assets pair					
Threat	Asset	Asset vulnerability (%)	Asset degradation (%)	Impact on asset (I) [€]	Intrinsic risk on asset (IR) [€]
01. Not finishing in time	1. TBM	20	0	=5,538,943 ×0=0	0
02. TBM malfunction	1. TBM	10	0	=5,538,943 ×0=0	0
03. Conveyor belt malfunction	5. Conveyor belt	5	0	=56,000 ×0=0	0

(continued)

Table 9.4 (continued)

Threat/assets pair					
Threat	Asset	Asset vulnerability (%)	Asset degradation (%)	Impact on asset (I) [€]	Intrinsic risk on asset (IR) [€]
04. Interruption of electrical energy for TBM	6. Electric generation plant	30	5	=128,564 ×0.05 =6,428	6,428×0.3 =1,928
05. Interruption of electrical energy for conveyor belt	5. Conveyor belt	30	3	= 56,000 ×0.03 =1,680	1,680×0.3 =504
06. Interruption of electrical concrete plant	3. Concrete plant	30	90	= 341,734×0.9 =307.560	307,560 ×0.3 =92,268
07. Different kind of soil	4. Spare cutting discs	50	50	= 10,845×0.5 =5,422	5,422×0.5 =2,711
08. Heavy filtrations coming from the river	1. TBM	25	20	=5,538,943 ×0.20 =1,107,788	1,107,788 ×0.25 =276,947
	2. Tunnel	20	18	1,586,207 ×0.18 =285,517	285,517 ×0.20 =57,103
				Total intrinsic risk:	431,461€

Table 9.5 Threats and safeguards pairs

Threat/Safeguards pair			
Threat	Safeguards	Vulnerability Decrease (VD) [%]	Impact Decrease (ID) [%]
01. Not finishing in time	001. Work extra time	60	0
02. TBM malfunction	002. Increase frequency maintenance periods	30	0
03. Conveyor belt malfunction	003. Increase frequency maintenance periods	20	0
04. Interruption of electrical energy for TBM	004. Look for another emergency source of electricity generation	0	95
05. Interruption of electric energy for conveyor belt	005. Have an emergency diesel generator	30	0
06. Interruption of electrical concrete plant	005. Have an emergency diesel generator	0	95
07. Different kind of soil	006. Increase frequency of inspections for cutting discs	20	0
08. Heavy filtrations coming from the river	007. Reinforcing the bed of the river	45	80

Safeguards are detailed in Table 9.5, and they are implemented using the mechanisms shown in Table 9.6. This table also depicts two columns, '*Implementation (IM)*', and '*Compliance (CM)*'

Compliance refers to a measurement of how suitable the safeguard mechanism is, and *Implementation* shows its performance level. That is, a mechanism may be very good in protecting something, but if for any reason it often ceases to operate, its performance can be qualified as low. For instance a handrail is very useful for preventing old people falling on a stair, so its compliance is high; however, if its anchors are not well embedded in a wall, the handrail becomes loose and no longer functions as a safeguard mechanism, therefore its implementation is low. Table 9.7 depicts the same Table 9.6 with added columns for computation of the effectiveness of each safeguard function.

Table 9.6 Safeguards and mechanism pairs

Safeguards/mechanisms pair			
Safeguards	Safeguard mechanisms	IM [%]	CM [%]
001. Work extra time	0001. Increase shift from 8 to 10 h	40	80
002. Increase frequency maintenance periods	0002. Hire a maintenance team besides the normal maintenance crew	60	80
003. Increase frequency maintenance periods	0003. Take advantage of TBM stops to also make maintenance of other equipment	30	90
004. Look for another emergency source of electricity generation	0004. Install a transformer to get electricity from the high voltage grid as an emergency	40	40
005. Have an emergency diesel generator	0005. Have a direct electrical line connected with the diesel and test its quick start every other day	100	80
005. Have an emergency diesel generator	0005. Have a direct electric line connected with the diesel power plant and test its quick star every other day	100	80
006. Increase frequency of inspections for cutting discs	0006. Develop with manufacturer a procedure to quickly replace discs in case of necessity	65	50
007. Reinforcing the bed of the river	0007. Pour a thick concrete slab on the bed of the river	90	100

Calculation of Effective Risk
The following formulas are used:

$$\text{Effective Vulnerability Decrease (EVD)} = \text{Vulnerability Decrease (VD)} \times \text{Effectiveness of Safeguards Function (EFe)}.$$

$$\text{Effective Impact Decrease (EID)} = \text{Impact Decrease (EID)} \times \text{Effectiveness of Safeguards Function (EFe)}.$$

$$\text{Effective Risk (ER)} = I(1 - \text{EID}) \times V(1 - \text{EVD}).$$

Table 9.7 Effectiveness computation of each safeguard function

Safeguards	Mechanisms	IM [%]	CM [%]	Effectiveness of safeguard functions (EFe)
001. Work extra time	0001. Increase shift from 8 to 10 h	40	80	0.4×0.8=0.32
002. Increase frequency maintenance periods	0002. Hire a maintenance team besides the normal maintenance crew	60	80	0.6×0.8=0.48
003. Increase frequency maintenance periods	0003. Take advantage of TBM stops to also make maintenance of other equipment	30	90	0.3×0.9=0.27
004. Look for another emergency source of electric generation	0004. Install a transformer to get electricity from the high voltage grid as an emergency	40	40	0.4×0.4=0.16
005. Have an emergency diesel generator	0005. Have a direct electrical line connected with the diesel and test its quick start every other day	100	80	1.00×0.8=0.8
005. Have an emergency diesel generator	0005. Have a direct electric line connected with the diesel power plant and test its quick start every other day	100	80	1.00×0.8=0.8
006. Increase frequency of inspections for cutting discs	0006. Develop with manufacturer a procedure to quickly replace discs in case of necessity	65	50	0.65×0.5=0.325
007. Reinforcing the bed of the river	0007. Pour a thick concrete slab on the bed of the river	90	100	0.9×1=0.9

See Table 9.8 for calculated Effective Risk.

It can be verified that the Effective Risk is lower than the Intrinsic Risk.

The ratio of Effective Risk and total assets is 0.035=3.5%.

This is the Residual Risk which is lower than the 8% risk willing to be assumed by the DM.

Table 9.8 Computation of effective risk

Threat	Safeguards	Vulnerability Decrease (VD) [%]	Impact Decrease (ID) [%]	EVD	EID	Effective Risk (ER) [€]
01. Not finishing in time	001. Work extra time	60	0	0.6×0.32=0.19	0	
02. TBM malfunction	002. Increase frequency maintenance periods	30	0	0.3×0.48=0.14	0	
03. Conveyor belt malfunction	003. Increase frequency maintenance periods	20	0	0.2×0.27=0.054	0	
04. Interruption of electrical energy for TBM	004. Look for another emergency source of electricity generation	0	95	0	0.95×0.16=0.152	1,635
05. Interruption of electrical energy for conveyor belt	005. Have an emergency diesel generator	30	0	0.30×0.8=0.24	0	383
06. Interruption of electric concrete plant	005. Have an emergency diesel generator	0	95	0	0.95×0.32=0.308	63,849
07. Different kind of soil	006. Increase frequency of inspections for cutting discs	20	0	0.2×0.32=0.065	0	2,534
08. Heavy filtrations coming from the river	007. Reinforcing bed of river	80	0	0.8×0.9=0.72		198,193
		Total effective risk:				**266,594**

9.2.2.5 Sequential Steps for Risk Detection, Evaluation and Safeguarding

1. Identify assets affected by diverse risks and compute their economic value. Establish a level of risk that the DM is willing to accept,
2. Find technical assets dependency and compute assets dependency value,
3. Determine potential risks or threats paired with assets and compute Intrinsic Risk (IR), that is the existing risk of doing nothing,
4. Determine safeguards to lessen or delete threats and pair them with safeguards,
5. Identify mechanisms, and pair safeguards with mechanisms,
6. Determine decrease in both vulnerability and impacts, because of mechanisms,
7. Compute Effective Risk (ER) that is risk after safeguard measures are considered,
8. Compare Effective Risk with Intrinsic Risk and compute the percentage in reduction. Remember that IR must be smaller than ER,
9. Compare this obtained risk percentage with the risk percentage that the DM was willing to accept. If the percentage lies below the accepted risk, the analysis succeeded in reducing it. The remaining risk percentage receives the name of Residual Risk (RR),
10. If there is no Residual Risk that is if the reduction is above the risk the DM is willing to accept, review the procedure and add more safeguard functions.

There is a family of software available to perform this calculation as well as a sensitivity analysis. Look for them on the Internet.

9.3 Outline of Actual Decision Problems Solved by Different Methods

Table 9.9 is a register of 66 projects that have been solved by different decision-making models. It is believed that it is useful information for consultation and also to learn about the different techniques that have been applied to work out diverse kinds of situations. Naturally, the results of this survey cannot prove that one method is better than any other; however, it is worth pointing out the abundance of projects treated with AHP, ELECTRE and PROMETHEE, which appear to be the leading methods, especially the first one.

Table 9.9 shows in its second column the original name of the project, as well as the area it belongs to, author or authors, source where the information has been obtained and corresponding electronic address, or particulars about other sources. The idea is to facilitate the interested reader's access to the original information where he/she can find all the aspects and details for each case.

As we can see, there are many projects solved by this writer using the Linear Programming or the SIMUS methods, and also employed by three researchers in their Ph.D. Thesis. This author's electronic address has been entered in each project solved by these two last methods, and the reader is encouraged to contact him either for consultation or to clarify some concepts.

Table 9.9 Actual projects and methods used to solve them

PROJECTS –TITLE –AREA –AUTHOR/S –SOURCES –ELECTRONIC ADDRESSES	Different models and methods used for solving actual cases								
	AHP ANP	ELECTRE	PROMETHEE	Cost/Benefit	Linear Programming and SIMUS	MAUT	Regime	SAMI	Other methods
01 ***Selection of river waterways – Interconnection*** RIVER WATERWAYS **Author:** Anonymous Source: Economic Development (132) page 123								X	
02 ***Integrated MCDA: A simulation case study*** AIRPORT TRAFFIC **Authors:** **Valerie Belton and Mark Elder** Source: http://www.simul8.ccom/support/library/mcda_sim.pdf									**Simulation** and **V.I.S.A.** **X**
03 ***A note on the selected multicriteria decision - Methods and their applications*** AIRPORT ROAD ACCESS **Authors:** **D·研 究 論 文 D** 홍 순 욱',조 s태 영동대학교 건설공학부 Graduate School of Business, University of Pittsburgh	**X**	**X**				**X**			**Goal Programing** **X**

(continued)

Table 9.9 (continued)

PROJECTS –TITLE –AREA –AUTHOR/S –SOURCES –ELECTRONIC ADDRESSES	Different models and methods used for solving actual cases AHP ANP	ELECTRE	PROMETHEE	Cost/Benefit	Linear Programming and SIMUS	MAUT	Regime	SAMI	Other methods
04 ***Cooperative games in infrastructure projects*** ROAD INFRASTRUCTURE **Author:** **Paola Pucci** Source: COST C8 - Best Practice in Sustainable Urban Infrastructure – International Conference, Trento Italy, 2003					**X**				
05 ***Analytic Network Process Model for highway corridor*** ROAD INFRASTRUCTURE **Author:** **Mongkut Piantanakulchai** Source: Sirindhorn International Institute of Technology http://ideas.repec.org/p/wpa/wuwpur/0509021.html	**X**								

06	***An integrated multi-criteria decision analysis and inexact mixed integer linear programming approach for solid waste management*** SOLID WASTE **Authors**: **S. Cheng** **C.W. Chang** **G.H. Huang** Source: Engineering Applications of Artificial Intelligence Volume 16, Issues 5-6, August-September 2003, Pages 543–554		X	**TOPSIS** X **CGT** X **SWA** X
07	***Determining the best blend for pavements*** ROAD INFRASTRUCTURE **Author:** **Tomasz Thiel** Source: http://www.liee.chemeng.ntua.gr/MCDA_53/PAPERS/paper_Tomasz%20Thiel%20PUT%20Poznan%20Poland.PDF		X	
08	***Selecting a car*** PURCHASING **Author:** Anonymous Source: http://www.google.es/search?hl=es&q=selecting+a+car+AHP+miles+per+gallon+0.123&meta=&aq=f&oq=	X		

(continued)

Table 9.9 (continued)

PROJECTS –TITLE –AREA –AUTHOR/S –SOURCES –ELECTRONIC ADDRESSES	Different models and methods used for solving actual cases AHP ANP	ELECTRE	PROMETHEE	Cost/Benefit	Linear Programming and SIMUS	MAUT	Regime	SAMI	Other methods
09 ***An Application of the Analytical Hierarchy Process to international location decision-making*** INDUSTRIAL LOCATION **Authors:** **Walailak Atthirawong** **Bart MacCarthy** Source**:** http://kipedia.org/wiki/Analytic_Hierarchy_Process	X								
10 ***A multicriteria decision support methodology for evaluating airport expansion plans*** AIRPORT POLICIES **Authors:** **Ron Vreeker** **Peter Nijkamp** **Chris Ter Welle** Source**:** http://www.tinbergen.nl/discussionpapers/01005.pdf							**X**		

11	***Multi-criteria planning of local energy with multiple energy carriers*** ELECTRICAL ENERGY SUPPLY **Author:** **Espen Loken** Source**:** Norwegian University of Science and Technology Faculty of Information Technology, Mathematics and Electrical Engineering Department of Electric Power Engineering http://www.dart-europe.eu/basic-search.php	**X**			**X**
12	***Ranking projects using the ELECTRE Method*** POWERHOUSE STRATEGIES **Authors:** **John Buchanan** **Paul Sheppard** Source: http://www.orsnz.org.nz/conf33/papers/p58.pdf		**X**		
13	***Medium hydro-power study project MHSP*** RIVER BASIN **Author:** Nepal Electricity Authority with technical cooperation from the Canadian International Water and Energy Consultants Source: http://www.south-asia.com/mhsp/mhsp.htm			**UNKNOWN, probably Cost/Benefit**	

(continued)

Table 9.9 (continued)

PROJECTS –TITLE –AREA –AUTHOR/S –SOURCES –ELECTRONIC ADDRESSES	Different models and methods used for solving actual cases								
	AHP ANP	ELECTRE	PROMETHEE	Cost/Benefit	Linear Programming and SIMUS	MAUT	Regime	SAMI	Other methods
14 ***A multicriteria approach for selecting a portfolio manager*** PEOPLE RECRUITING **Authors:** **Moez Hababou,** **Jean-Marc Martel** Source: Special issue of INFOR on performance analysis, August 1998 http://moezh.tripod.com/paper.pdf			X						
15 ***Some particular aspects concerning ELECTRE methods applications*** INVESTMENT ALTERNATIVES **Author:** **Gheorge Condurache** **Romeo-Mihai Ciobanu** Source: http://www.unitus.it/mcda57/lavori/PapConduracheCiobanu.pdf		X							

16	***Dwellings construction in Geneva – Switzerland*** HOUSING **Author:** **MUTATE** Source: *Ref: http://ecolu-info.unige.ch/~haurie/mut a1te/Mutate_final/Lectures/Lect_1_3_2/ lect_1_3_2.htm#toc*	X	
17	***Water distribution – Brazil*** WATER DISTRIBUTION **Authors:** **Danielle Morais** **Adiel Almeida** Source: http://www.wrc.org.za/downloads/ watersa/2006/April%2006/1869.pdf	X	
18	***Selection between cesspools or public sewage- Oslo, Norway*** SEWAGE **Authors:** **Oddvar Gregg Lindholm** **Kine Halvorsen Thoren** Source: www.cf.ac.uk/research/ bost8/case/watersewerage/ conventionalnaturesewerage.html		X

(continued)

Table 9.9 (continued)

PROJECTS	Different models and methods used for solving actual cases								
–TITLE –AREA –AUTHOR/S –SOURCES –ELECTRONIC ADDRESSES	AHP ANP	ELECTRE	PROMETHEE	Cost/Benefit	Linear Programming and SIMUS	MAUT	Regime	SAMI	Other methods
19 ***PROMETHEE multicriteria analysis for evaluation of recycling strategies in Malaysia*** RECYCLING STRATEGIES **Authors:** **Santha Chenayah** **Eiji Takeda** Source: http://ideas.repec.org/p/osk/wpaper/0501.html			X						
20 ***PROMETHEE multicriteria analysis for evaluation of recycling strategies in Malaysia*** RECYCLING STRATEGIES **Authors:** **Santha Chenayah** **Eiji Takeda** Source: http://ideas.repec.org/p/osk/wpaper/0501html			X						

21	***Aplicaçao conjunta de métodos de apoio multicriterio*** LOCATION ANALYSIS **Author:** **Autrán Monteiro Gomes** Source: Revista argentina de I.O, año XI. No 23, pag. 69 to 86, Nov.2002			X
22	***Multi-criteria analysis manual*** GOVERNMENT **Author:** **Department for Transport, Local Government and the Regions DTLR** Source: http://www.sfu.ca/mpp/pdf_news/811-04%20UK%20MCA%20Manual.pdf		X	
23	***The MCDM based redesign of the distribution system – Poland*** MERCHANDISE DISTRIBUTION **Authors:** **Jack Zak** **Hanna Wlodarczak** **Marcin Kicinski** Source: http://www.iasi.cnr.it/ewgt/13conference/119_zak.pdf	X	X	

(continued)

Table 9.9 (continued)

	PROJECTS –TITLE –AREA –AUTHOR/S –SOURCES –ELECTRONIC ADDRESSES	AHP ANP	ELECTRE	PROMETHEE	Cost/Benefit	Linear Programming and SIMUS	MAUT	Regime	SAMI	Other methods
		Different models and methods used for solving actual cases								
24	***Baku-Tbilisi-Ceyhan pipeline*** OIL PIPELINE **Data from:** *http://en.wikipedia.org/wiki/Baku%E2%80%93Tbilisi%E2%80%93Ceyhan_pipeline* **Solved by:** **Nolberto Munier** nolmunier@yahoo.com				**UNKNOWN – PROBABLY INTERNATIONAL ASPECTS AND POLITICAL CIRCUMSTANCES WERE CONSIDERED**					
25	***Example: Three projects to prioritize*** AGRICULTURE **Author:** **University of Iowa**	**X**								

26	***Use of geographic information systems in an environmental impact assessment of an overhead power line*** ENERGY TRANSMISSION **Authors:** **L. Warner** **R.D. Diab** Source: http://www.ingentaconnect.com/content/beech/iapa/2002/00000020/00000001/art00005		**GIS X**
27	***Multicriteria choice of ore transport system to an underground mine: application of PROMETHEE methods*** MINING **Authors:** **B. Elevli** **A. Demirci** Source : http://www.saimm.co.za/publications/downloads/v104n05p251.pdf	**X**	

(continued)

Table 9.9 (continued)

PROJECTS –TITLE –AREA –AUTHOR/S –SOURCES –ELECTRONIC ADDRESSES	Different models and methods used for solving actual cases								
	AHP ANP	ELECTRE	PROMETHEE	Cost/Benefit	Linear Programming and SIMUS	MAUT	Regime	SAMI	Other methods
28 ***Multicriteria decision-making using ELECTRE*** LAND USE **Author:** **Anonymus** Source: http://www.easts.info/on-line/proceedings_05/2237.pdf		**X**							
29 ***Multicriteria methods*** **Author:** **Evangelos Triantaphyllou** Multi-Criteria Decision Making Methods – Kluwer Academics Publishers	**X**								

30	***Strategic implementation of infrastructure priority projects: Case study in Palestine*** REHABILITATION **Authors:** **K. Nigim** **M. Ziara** Source: http://www.ece.uwaterloo.ca/~knigim/Ziara&Nigim-ASCE-infrastructure_March_2002.pdf	**X**		
31	***Transport-sustainable urban development strategies*** URBAN TRANSPORTATION **Data from authors:** **Antonio Ferreira** **Antonio Antunes** **Paulo Pinho** **Data from:** COST8 Final Conference **Solved by:** **Nolberto Munier** nolmunier@yahoo.com		**X**	**Unknown**
32	***Urban constructions*** HOUSING DEVELOPMENT **Author:** **Nolberto Munier** nolmunier@yahoo.com		**X**	

(continued)

Table 9.9 (continued)

PROJECTS –TITLE –AREA –AUTHOR/S –SOURCES –ELECTRONIC ADDRESSES	Different models and methods used for solving actual cases								
	AHP ANP	ELECTRE	PROMETHEE	Cost/Benefit	Linear Programming and SIMUS	MAUT	Regime	SAMI	Other methods
33 ***Strategic environmental assessment of Canadian energy policy*** ENERGY POLICIES **Author:** **Bram F. Noble** Source: Impact Assessment and Project Appraisal, September 2002			**UNKNOWN**						
34 ***Environmental indicators selection*** ENVIRONMENT **Author:** **Nolberto Munier** nolmunier@yahoo.com					X				
35 ***Identification of possible options for a policy to promote recycling in the paper industry*** RECYCLING **Author:** **Nolberto Munier** nolmunier@yahoo.com					X				

36	***Selection of urban indicators*** ENVIRONMENT **Author:** **Nolberto Munier** nolmunier@yahoo.com			X
37	***Selection of site for waste incinerators*** LOCATION ANALYSIS **Author:** **Laura Tasca** Source: Ph.D.Thesis – University of Milano, Italy		X	X
38	***Urban projects*** REHABILITATION **Author:** **Nolberto Munier** nolmunier@yahoo.com			X
39	***Priority polices in agriculture*** AGRICULTURE **Authors:** **Mina ĐORĐEVIÆ1** **Tihomir Hunjak** Source: Agriculturae Conspectus Scientificus, Vol. 67 2002) No. 2 (57-68)	X		

(continued)

Table 9.9 (continued)

PROJECTS	Different models and methods used for solving actual cases								
–TITLE –AREA –AUTHOR/S –SOURCES –ELECTRONIC ADDRESSES	AHP ANP	ELECTRE	PROMETHEE	Cost/Benefit	Linear Programming and SIMUS	MAUT	Regime	SAMI	Other methods
40 ***Selection of urban and periurban indicators*** ENVIRONMENT **Author:** **Rocío Díaz-Chaves** Source : Ph.D. Thesis Aberyswyth University-UK					**X**				
41 ***Multicriteria evaluation of sustainable agricultural land use. a case study of Lesvos*** AGRICULTURE AND CATTLE **Authors:** **Cerda Hermanides** **Peter Nijkamp** Source: Multicriteria Analysis for Land-Use Management, 61-78 Research Memorandum 1997-5, Dept of Economics, Free University, Amsterdam, 1997 http://hdl.handle.net/1871/3356 Nijkamp, prof. dr. P.							**X**		

42 ***Guidelines in formulating city development strategies – Capital investments*** URBAN INVESTMENTS **Data:** **From unknown author** **Solved by:** **Nolberto Munier** nolmunier@yahoo.com		X
43 ***Selection of urban road alternatives*** URBAN INFRASTRUCTURE **Author:** **Nolberto Munier** nolmunier@yahoo.com		X
44 ***Environmental impact evaluation*** ENVIRONMENT **Author:** **Nolberto Munier** nolmunier@yahoo.com		X
45 ***Una aplicación de ELECTRE a la selección de un caza-bombardero*** AIR FORCE **Author:** **Carlos Romero** Source: Análisis de las decisiones multicriterio – ISDEFE, Madrid, 1996	X	

(continued)

Table 9.9 (continued)

PROJECTS –TITLE –AREA –AUTHOR/S –SOURCES –ELECTRONIC ADDRESSES	Different models and methods used for solving actual cases AHP ANP	ELECTRE	PROMETHEE	Cost/Benefit	Linear Programming and SIMUS	MAUT	Regime	SAMI	Other methods
46 ***Multi-criteria decision support for integrated technique assessment*** INDUSTRY **Authors:** **Martin Treitz, Hannes Schollenberger, Benjamin Schrader, Jutta Geldermann, Otto Rentz** RadTech Europe 2005 Conference & Exhibition Source: http://www.radtech europe.com/files_content/march%202005%20papers/treitzpapermarch2006.pdf			X						

47	***Strict uncertainty*: *A criterion for moderately pessimistic decision maker* THEORY *to rank actions under uncertainty* **Author:** **Enrique Ballestero** Source: Decision Sciences: winter 2002, 33;1; ABI/INFORM Global	**Compromise Programming** **X**
48	***Instrumentos económicos para La gestión ambiental*: *Decisiones monocriteriales vs. Decisiones multicriteriales*** FORESTRY **Authors:** **F. Falconi** **R. Burano** Source: Revista Iberoamericana de Economía Ecológica Vol. 1: 11–20 http://www.redibec.org/archivos/revista/articulo2.pdf	**NAIADE**

(continued)

Table 9.9 (continued)

PROJECTS –TITLE –AREA –AUTHOR/S –SOURCES –ELECTRONIC ADDRESSES	Different models and methods used for solving actual cases AHP ANP	ELECTRE	PROMETHEE	Cost/Benefit	Linear Programming and SIMUS	MAUT	Regime	SAMI	Other methods
49 ***Pre-feasibility MCDM tools to aid communities in priorizing viable renewable energy sources*** RENEWABLE ENERGY **Authors:** **K. Nigim** **J. Green** **N. Munier** Source: Renewable Energy 29 (2004) 1775–1791	**X**								
50 ***Localización de planta de generación eléctrica en Europa*** INDUSTRIAL LOCATION **Author:** **Pablo Aragonés Beltrán** Source: Universidad Politécnica de Valencia, Spain			**X**						

51 ***Decision analysis*** MEDICINE **Author:** **Posavac & Carey, (1989).** Source: http://www.fammed.ouhsc.edu/tutor/decanal.htm		**X**
52 ***Community infrastructure upgrading for Ghana*** HOUSING **Data**: World Bank publication '*Toward a National Community Infrastructure Upgrading Program for Ghana*' Solved by: **Nolberto Munier** nolmunier@yahoo.com	**X**	
53 ***Urban rehabilitation*** URBAN REHABILITATION **Author:** **Salah Taha** Source: Ph.D. Thesis, University of Gaza, Palestine	**X**	
54 ***Selecting urban intersections*** URBAN STRATEGY **Author:** **Nolberto Munier** nolmunier@yahoo.com	**X**	

(continued)

Table 9.9 (continued)

PROJECTS –TITLE –AREA –AUTHOR/S –SOURCES –ELECTRONIC ADDRESSES	Different models and methods used for solving actual cases AHP ANP	ELECTRE	PROMETHEE	Cost/Benefit	Linear Programming and SIMUS	MAUT	Regime	SAMI	Other methods
55 ***Community development*** URBAN STRATEGY **Author:** **Nolberto Munier** nolmunier@yahoo.com					**X**				
56 ***Selection measures to decrease traffic accidents*** URBAN STRATEGY **Author:** **Nolberto Munier** nolmunier@yahoo.com					**X**				
57 ***Selection of urban sustainable indicators*** ENVIRONMENT **Author:** **Nolberto Munier** nolmunier@yahoo.com					**X**				
58 ***Determining spatial location for urban health centres*** URBAN STRATEGY **Nolberto Munier** nolmunier@yahoo.com					**X**				

59	***Selection of urban indicators*** URBAN STRATREGY **Data from Leon City Hall Mexico** Solved by**:** **Nolberto Munier** nolmunier@yahoo.com	**X**
60	***Selection of public works (Sewage)*** URBAN STRATEGY **Data from Córdoba City Hall (Argentina)** Solved by: **Nolberto Munier** nolmunier@yahoo.com	**X**
61	***Scheduling bridges repairs*** REGIONAL INFRASTRUCTURE **Data from Clarence-Rockland City Hall (Canada)** Solved by: **Nolberto Munier** nolmunier@yahoo.com	**X**
62	***Sustainable indicators*** ENVIRONMENT **Data from:** **Guadalajara City Hall and** **Monterrey Polytechnic (México)** Solved by: **Nolberto Munier** nolmunier@yahoo.com	**X**

(continued)

Table 9.9 (continued)

PROJECTS –TITLE –AREA –AUTHOR/S –SOURCES –ELECTRONIC ADDRESSES	Different models and methods used for solving actual cases AHP ANP	ELECTRE	PROMETHEE	Cost/Benefit	Linear Programming and SIMUS	MAUT	Regime	SAMI	Other methods
63 ***Business location case study*** COMMERCE **Author:** **Valerie Belton** **Theodor Stewart** Source: Management Science, University of Strathclyde, Glasgow, Scotland1					**UNKNOWN**				
64 ***Selección de una tecnología de banda ancha para la Universidad de Colombia*** COMMUNICATIONS **Authors:** **Félix Antonio Cortez Aldana** **Mónica García Melón** **Pablo Aragonés** Source: Ing.Investig. vol.27 no.1 Bogotá Jan./Apr. 2007	**X**								

65 ***A multicritertia approach to decision aid*: *PROMETHEE & GAIA*** ADVERTISEMENT **Author:** **Bertrand Marechal** Source: http://theses.ulb.ac.be/ETD-db/collection/available/ULBetd-06162008-172523/unrestricted/Annexe_PROMETHEE.pdf	**X**
66 ***An integrated multi-criteria system to assess sustainable energy options*: *An application of the PROMETHEE method*** ENERGY **Author:** **Fausto Cavallaro** Source: http://www.feem.it/NR/rdonlyres/9E940C73-A958-4D9C-B5CC-48E940D89829/1448/2205.pdf	**X**

References

Aragonés, P. (2008). *Análisis Multicriterio de decisiones. Métodos PROMETHEE*. Valencia: Apuntes de clase, Universidad Politécnica de Valencia.

Belton, V., & Stewart, T. (2002). *Multiple criteria decision analysis – An integrated approach*. Boston: Kluwer.

Fernández, M., & Munier, N. (2010). *Bases para la gestión de riesgos en proyectos*. València: Editoral Universitat Politècnica de València.

Löken, E. (2007). *Multi-criteria planning of local energy systems with multiple energy carriers*. Thesis for the degree Philosophy Doctor, Norwegian University of Science and Technology, Trondheim.

Morais, D., & Almeida, A. (2006). Water supply system decision making using multicriteria analysis. *African Journals on Line-Water SA, 32*(2), 229–236.

Munier, N. (2004). *Multicriteria environmental assessment – A practical guide*. Dordrecht: Kluwer.

Glossary

Aalborg's 10 Commitments Series of the compromises signed in the Danish city of this name with the purpose to improve urban sustainability in each village, town and city.

Agenda 21 International regulation that encourages municipalities to become sustainable. This is an initiative directly derived from the Earth Summit in Rio de Janeiro.

Attribute Proper characteristic of an object, for instance: 'luminous', 'soft', 'elegant', etc.

Contribution See 'score'.

Criterion Fundamental element that establishes guidelines that the diverse alternatives must comply.

Decision matrix Table or matrix, which contains for each criterion the scores corresponding to each alternative.

Decision table Table with the set of scores of diverse alternatives in relationship with criteria.

Earth Summit Refers to the Convention on Global Diversity held in Rio de Janeiro in 1992. This Convention attended by 152 world leaders led to the signature of diverse agreements on biological diversity and on many sustainable aspects including Agenda 21.

Ecology Science that investigates the relationships between organisms and their environment.

Economy of scale When the production unit cost decreases along with a production increase.

Eigenvalue Scalar value associated with a linear transformation which, when applied to the transformation, changes it.

Evaluation To establish the value of an action, for instance assess the damage done by a specific impact.

Expert opinion or expert judgement Estimation, view, attitude or judgement from persons knowledgeable or skilled on a certain matter. For instance, the maximum value of the rate used when extracting water from a well and measured in m^3/s, or minimum floor space in a house measured in m^2.

N. Munier, *A Strategy for Using Multicriteria Analysis in Decision-Making: A Guide for Simple and Complex Environmental Projects*, DOI 10.1007/978-94-007-1512-7,

Impact In this context, it is the change of a condition of the ecosystem, in society or in the economy, caused by the development and implementation of a project.

Independent term In an equation or inequality is the term at the right of the corresponding operator sign.

Indicators Values or metrics that measure or gauge determined aspects of a project. For instance, in a water distribution system, an indicator is the estimated amount of water that a persons needs per day. Another is the Gross Domestic Product per inhabitant.

Internal Rate of Return Percentage on the investment, which is normally used to gauge the profitability of a project.

Lagrange Multiplier It is equivalent to the marginal value in Linear Programming, that is, it indicates how the objective function changes with the unit variation in a criterion. The Lagrange multiplier measures the instantaneous variation in Non-Linear Programming, i.e. when the objective function is not a straight line but a curve when it is tangent to the solution polytope.

Landfill (of domestic wastes) Lot of soil designated to storage wastes.

Lineal transformation Mathematical function that operates between vector spaces. Given a vector in an original vector space, it finds the correspondence of that vector space using additive operations and scalar multiplication.

MADM Acronym of Multi Attribute Decision Making, which is the decision-making process when several attributes are considered for evaluation of alternatives.

Marginal value Magnitude that indicates how much a solution changes when a criterion threshold varies by one unit.

Model A mathematical representation, generally imperfect, of an actual system. For instance, a regression model between two variables tries to replicate the existent conditions in an actual case where two variables play, and determines how one variable changes with respect to the other, although it does not necessarily indicate a cause and effect relationship. In Economics, for instance, a classical example of a model is the Input-output matrix, which relates the industrial activities in a country.

MODM Acronym for Multi Objectives Decision Making, which is the decision-making process when several objectives are simultaneously considered.

Net Present Value Quantity of monetary units that measure the net economic output of a project, when its estimated net results along several years are brought to present values.

NOx General formulation to indicate chemical compounds formed by nitrogen and oxygen.

Objective function Mathematical function that expresses the goal of a problem as a function of the intervening variables.

Objective Goal, aspiration, desire, that is something that we want to reach or obtain.

Objectives values Amounts that do not respond to opinions but to actual facts, for instance the production of vehicles in a car manufacturing plant.

Paired comparison matrix Given a criteria set, a paired comparison evaluates a criterion for each one of the others, which results form a matrix.

Required value It is the independent term or threshold. For instance, when it is demanded that the economic value of all projects must not exceed a certain money amount, which is the required and maximum quantity available.

Restriction Criterion limited by a threshold, that is, the independent term.

Score In the decision table, it is a value placed at the intersection of a column corresponding to an alternative and a row related to a criterion. It expresses the contribution of each alternative according to the criterion requirement.

Shadow price Marginal value for a criterion.

SOx General formulation to indicate chemical compounds formed by sulphur and oxygen, usually produced by the combustion of fuel containing sulphur.

Stakeholders Interested parties, that is, people who can be affected by the impacts of a project, i.e. government, decision entities, community associations, industry and public in general.

Subjective values Amounts that respond to opinions and personal judgments, for instance behavioural expressions such as 'good', 'very good', 'interesting', 'pretty', etc.

To estimate To render a value judgement or opinion with reference to the value of something.

To normalize: Unify This process is applied to a decision matrix when there are dissimilar values. Normalization produces magnitudes between 0 and 1 which can be compared.

Utility function It is determined individually for each alternative and is equal to the product of its score by the weight assigned to the criterion where the score belongs.

Utility It has several meanings according to its application. In multicriteria theory, it expresses the degree of satisfaction that a user obtains from a certain product or service.

Weight Degree of importance assigned to projects, alternatives and criteria.

Index

MIX
Papier aus verantwortungsvollen Quellen
Paper from responsible sources
FSC® C105338

If you have any concerns about our products,
you can contact us on
ProductSafety@springernature.com

In case Publisher is established outside the EU,
the EU authorized representative is:
Springer Nature Customer Service Center GmbH
Europaplatz 3, 69115 Heidelberg, Germany

Printed by Libri Plureos GmbH
in Hamburg, Germany